T0234733

SpringerBriefs in Applied Sciences and Technology

Manufacturing and Surface Engineering

Series Editor

Joao Paulo Davim⬤, Department of Mechanical Engineering, University of Aveiro, Aveiro, Portugal

This series fosters information exchange and discussion on all aspects of manufacturing and surface engineering for modern industry. This series focuses on manufacturing with emphasis in machining and forming technologies, including traditional machining (turning, milling, drilling, etc.), non-traditional machining (EDM, USM, LAM, etc.), abrasive machining, hard part machining, high speed machining, high efficiency machining, micromachining, internet-based machining, metal casting, joining, powder metallurgy, extrusion, forging, rolling, drawing, sheet metal forming, microforming, hydroforming, thermoforming, incremental forming, plastics/composites processing, ceramic processing, hybrid processes (thermal, plasma, chemical and electrical energy assisted methods), etc. The manufacturability of all materials will be considered, including metals, polymers, ceramics, composites, biomaterials, nanomaterials, etc. The series covers the full range of surface engineering aspects such as surface metrology, surface integrity, contact mechanics, friction and wear, lubrication and lubricants, coatings an surface treatments, multiscale tribology including biomedical systems and manufacturing processes. Moreover, the series covers the computational methods and optimization techniques applied in manufacturing and surface engineering. Contributions to this book series are welcome on all subjects of manufacturing and surface engineering. Especially welcome are books that pioneer new research directions, raise new questions and new possibilities, or examine old problems from a new angle. To submit a proposal or request further information, please contact Dr. Mayra Castro, Publishing Editor Applied Sciences, via mayra.castro@springer.com or Professor J. Paulo Davim, Book Series Editor, via pdavim@ua.pt.

More information about this subseries at http://www.springer.com/series/10623

Marcel Kuruc

Rotary Ultrasonic Machining

Application for Cutting Edge Preparation

 Springer

Marcel Kuruc
Slovak University of Technology in Bratislava
Bratislava, Slovakia

ISSN 2191-530X   ISSN 2191-5318 (electronic)
SpringerBriefs in Applied Sciences and Technology
ISSN 2365-8223   ISSN 2365-8231 (electronic)
Manufacturing and Surface Engineering
ISBN 978-3-030-67943-9   ISBN 978-3-030-67944-6 (eBook)
https://doi.org/10.1007/978-3-030-67944-6

© The Author(s), under exclusive license to Springer Nature Switzerland AG 2021
This work is subject to copyright. All rights are solely and exclusively licensed by the Publisher, whether the whole or part of the material is concerned, specifically the rights of translation, reprinting, reuse of illustrations, recitation, broadcasting, reproduction on microfilms or in any other physical way, and transmission or information storage and retrieval, electronic adaptation, computer software, or by similar or dissimilar methodology now known or hereafter developed.
The use of general descriptive names, registered names, trademarks, service marks, etc. in this publication does not imply, even in the absence of a specific statement, that such names are exempt from the relevant protective laws and regulations and therefore free for general use.
The publisher, the authors and the editors are safe to assume that the advice and information in this book are believed to be true and accurate at the date of publication. Neither the publisher nor the authors or the editors give a warranty, expressed or implied, with respect to the material contained herein or for any errors or omissions that may have been made. The publisher remains neutral with regard to jurisdictional claims in published maps and institutional affiliations.

This Springer imprint is published by the registered company Springer Nature Switzerland AG
The registered company address is: Gewerbestrasse 11, 6330 Cham, Switzerland

Preface

Efficiency of the machining process is a key attribute for increasing the productivity and reducing the costs of the process. It could be achieved by several ways: by developing the advanced machining methods; by developing the new cutting materials; by modification of the cutting tool design; by creating advanced coatings on the cutting tool; by optimizing the cutting parameters, etc.

Issues with efficiency concerns especially machining of newly developed advanced materials. Those materials usually have to meet unique properties, such as high strength, high hardness, high toughness, elevated heat resistance and elevated creep resistance. And those unique properties can worsen machinability. Machining of those materials by common technologies was ineffective, expensive and sometimes even impossible. For example, machining of hardened steel by high-speed steel (HSS) cutting tool is almost impossible. When carbide cutting tool is used, only very low cutting speeds are recommended, which makes the process ineffective. When proper coating is used, efficiency of the process will arise; however, there is still a rapid tool wear present. The highest efficiency is achieved when cubic boron nitride (CBN) cutting inserts are used. This cutting material is not cheap; therefore, its highest possible tool life is requested. It could be reached by cutting-edge preparation.

Cutting-edge preparation is based on controlled modification (fillet, chamfer) of cutting edge. However, processes which are possible to manufacture such hard, heat resistant and electric isolator material as CBN are few. One of the machining methods, which were developed to machining very hard and brittle materials, is rotary ultrasonic machining (RUM). RUM is relatively new technology, which has a big potential in many industrial sectors, as well as in medicine. The results reported in this scientific monograph could be implemented for optimization of the cutting parameters of RUM process during machining of the advanced ceramic materials, as well as for successful cutting-edge preparation of CBN inserts.

Trnava, Slovakia   Marcel Kuruc

Acknowledgements

This scientific monograph was written with support of the project of VEGA grant agency of the Ministry of Education, Science, Research and Sport of the Slovak Republic and Slovak Academy of Sciences, no. 1/0097/17: "The research of novel method for cutting edge preparation to increase the tool performance in machining of difficult-to-machine materials"; and with the support of the APVV Project of Slovak Research and Development Agency of the Ministry of Education, Science, Research and Sport of the Slovak Republic, no. APVV-16-0057: "Research into the Unique Method for Treatment of Cutting Edge Microgeometry by Plasma Discharges in Electrolyte to Increase the Tool Life of Cutting Tools in Machining of Difficult-to-Machine Materials."

Summary

In the present, there grows the interest in advanced materials, due to their specific properties, such as high strength, high hardness, high abrasion resistance, high toughness, high thermal and chemical stability, high corrosion resistance, low mass and other significant properties. However, these unique properties of theirs cause problems during their manufacturing. Advanced methods of machining have been created to treat these types of materials. The ultrasonic machining belongs to them. Usage of ultrasound enhances almost every industrial sector. During machining itself, it increases the accuracy of the machined surface, elongates tool life, decreases cutting forces and the generation of process heat, reduces surface roughness and allows to machine hard and brittle materials mechanically, since the machining of such materials by any other way is quite problematic. A lot of advanced materials are produced in the basic defined shapes due to the difficulties of their processing. However, there exist applications, where their shapes have to be more complex, which standard methods of production of these materials cannot fulfil. Rotary ultrasonic machining is able to machine such materials. Different shapes and sizes of the ultrasonic tools allow to produce complex shapes of products, which cannot be achieved by grinding or material production process.

This scientific monograph deals with the issues of the process and conditions of rotary ultrasonic machining (RUM) of hard materials and the recommendation (according to performed researches) of proper machining parameters (for selected hard materials). Then, the obtained machining parameters were applied for cutting-edge preparation of CBN cutting inserts. There were observed suitability of RUM for creation of chamfers on the cutting edge, as well as measured accuracy of the process. In the experiments, the size and the angle of the chamfer were changing. According to the obtained results of the analysis of the cutting edge, it can be concluded: RUM is able to create controlled preparation; however, dimensions did not meet expected accuracy. There were determined deviation of achieved dimensions, created influences of selected deviations, calculated obtained depth differences and identified possible sources of observed inaccuracies.

The monograph can serve as guidebook of rotary ultrasonic machining with its practical implementation for cutting-edge preparation with highlighted possible issues, which can occur during this process, as well as solutions how to solve them.

Contents

About the Author

Marcel Kuruc, M.Sc., Ph.D. works for the Slovak University of Technology in Bratislava, Faculty of Materials Science and Technology in Trnava, where he teaches the subjects: Introduction to Computer-Aided Production Technologies; Computer-Aided Production Technologies I; CAD/CAM; Computer-Aided Technologies and Systems; Computer-Aided Assembly; Theory of Machining; Technology of Machining and Assembly; Technology of Assembly; Advanced Machining Methods. His professional interests are generally directed to the rotary ultrasonic machining; laser beam machining; computer-aided design (CAD); computer-aided manufacturing (CAM); reverse engineering; multi-axis machining; CNC programming. He participated in a few teacher mobilities and attended numerous international conferences in many countries. He is an author of several professional and scientific articles, as well as a few patents. He has been a member of several research teams dealing with scientific projects (APVV, VEGA, KEGA, H2020) in the field of machining and related fields. In addition, he finds it important to disseminate his knowledge and experience in order to promote science and technology to the students at secondary schools and universities, as well as to industrial companies.

Department of Machining and Computer Aided Technologies
Institute of Production Technologies
Faculty of Materials Science and Technology in Trnava
Slovak University of Technology in Bratislava
Trnava, Slovak Republic
e-mail: marcel.kuruc@stuba.sk

Nomenclature

AC	Alternating current
BUE	Build up edge
CBN	Cubic boron nitride
CFRP	Carbon fibre-reinforced plastic
CMC	Ceramic matrix composite
DG	Diamond grinding
DOC	Depth of cut (a_p)
EDM	Electrical discharge machining
EDX	Energy-dispersive X-ray spectroscopy
FEM	Finite element method
FFS	Free-form surface
H/E	Hardness to Young's modulus ratio
HPHT	High pressure, high temperature
HSS	High-speed steel
KPD	Kalium phosphorus dihydrogen—potassium dihydrogen phosphate
LASER	Light amplification by stimulated emission of radiation
LBM	Laser beam machining
LED	Light-emitting diode
MCD	Monocrystalline diamond
MMC	Metal matrix composite
MRR	Material removal rate
PCBN	Polycrystalline boron nitride
PCD	Polycrystalline diamond
rpm	Revolutions per minute
RUM	Rotary ultrasonic machining
SD	Synthetic diamond
SS	Stainless steel
US	Ultrasound, ultrasonic
USM	Ultrasonic machining
WJM	Water jet machining

Chapter 1
Introduction

In present, there grows interest in advanced materials, due to their specific properties, such as high strength, high hardness, high abrasion resistance, high toughness, high thermal and chemical stability, corrosion resistance, low mass and other significant properties. However, these unique properties of theirs cause problems during their manufacturing. Advanced methods of machining have been created to treat these types of materials. The ultrasonic machining belongs to them.

Usage of ultrasound enhances almost every industrial sector. During machining itself, it increases the accuracy of the machined surface, elongates tool life, decreases cutting forces and the generation of process heat, reduces the value of surface roughness parameters and allows to machine hard and brittle materials mechanically, since the machining of such materials by any other way is quite problematic. Technical glass and ceramics are the most common materials for ultrasonic machining.

Ultrasound can be used also to assist in the machining of materials, which are considerably softer and tougher, such as various metal alloys. It could improve the machining process by achieving better cutting conditions, and therefore, better parameters of the machined surface could be reached.

Higher accuracy of machined parts is required, due to the development of modern products. These products demand both high precision and complex shape. However, their parameters are limited by the development of machine tools. Therefore, there were many types of ultrasonic machine tools created, such as rotary ultrasonic milling machine tool. These machine tools are often combined with additional advanced principles, like operating in five axes, or high-speed spindle. Usage of continuous five axes machining allows processing complex shapes of component, which is a great benefit in present. High-spindle speed is utilized for machining by tools with a small diameter to keep recommended cutting conditions even when fine features are requested to be created. Ultrasonic energy enables to machine almost every material of workpiece and increases the quality of the surface of machined components as well.

Therefore, rotary ultrasonic machining could be used to manufacture complex components made of advanced materials, or modify or repair components made of

© The Author(s), under exclusive license to Springer Nature Switzerland AG 2021
M. Kuruc, *Rotary Ultrasonic Machining*, Manufacturing and Surface Engineering,
https://doi.org/10.1007/978-3-030-67944-6_1

hard materials, which should be adapted to new conditions, or renew to keep suitable properties in initial conditions.

Chapter 2
History of the Ultrasound

Abstract The inception of the industrial application of the ultrasound was conditional of several discoveries and inventions. The human ear cannot hear the ultrasound (it is practically its definition), therefore for the initial stimulus can be considered the discovery, there can exist the sound beyond audible frequency. After that there was a challenge to invent the device, which can produce the ultrasound. When such a device was manufactured, and when ultrasound could be modulated in different frequencies and amplitudes, the researchers could start thinking about its industrial application. The first global usage of the ultrasound in the industry was a military application in the detection of the submarines and aircraft. After World War II, there begin application of the ultrasound for peaceful purpose in the industry and medicine.

The first knowledge about sound out of the audible range (audible range of sound: 20 to 20,000 Hz) was discovered by Italian physiologist Lazzaro Spallanzani in 1793. He was examining navigation of bats in darkness. He made the bats blind, and he found out that bats can navigate themselves without any problem. He reached a decision that they were utilizing (ultra)sound to navigate, but he was not able to prove it [12]. Nowadays, it is generally known that for orientation in darkness the bats utilize ultrasonic vibrations with high frequencies (about 40 kHz), which is reflected from barriers, as shown in Fig. 2.1. Thanks to this echolocation, they can detect, localize and even classify various objects in complete darkness.

Almost one hundred years later, a German physicist Rudolph Koenig made, inter alia, whistles to find the highest hearable sound. His devices were able to generate sound in the frequency range from 4 to 90 kHz [6]. In 1883, sir Francis Galton took results of Koenig and made a whistle known as Galton whistle. It is an ultrasonic whistle which produces sound with frequency from 23 to 54 kHz, and it is utilized in dog training [3]. Since the end of the nineteenth century, more researches dealt with the properties of ultrasound at the theoretical level; however, none of them had any idea what to do with the ultrasound, which they created for practical reasons [1, 13].

One of the first utilizations for ultrasound was done in France during World War I. Professor Paul Langevin created high-power ultrasonic generator based on siliceous crystals. He put this device under water and found out that small fish which was under the generator died [7]. He was the first person, who used siliceous crystals to

© The Author(s), under exclusive license to Springer Nature Switzerland AG 2021
M. Kuruc, *Rotary Ultrasonic Machining*, Manufacturing and Surface Engineering,
https://doi.org/10.1007/978-3-030-67944-6_2

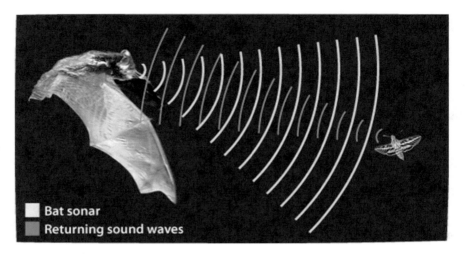

Fig. 2.1 Bat's echolocation [4]

generate high-frequency vibration. This effect is known as inverted piezoelectricity [13]. Piezoelectricity has been discovered by Pierre Currie in 1877 (Langevin was his doctoral student). At the end of World War I (1917), Langevin in cooperation with Constantin Chilowsky invented a hydrophone-sonar (ultrasound device) for underwater detection of hostile submarines. The first registration of a patent for an acoustic echo-locator in air, as well as in water, was performed by Lewis Fry Richardson in 1912 [8].

The first article, which described the possibility of using ultrasound for the machining, was published in 1927 by Wood and Loomis [9, 15]. However, the first patent for industrial usage of ultrasound was filed in 1945 by Professor William Balamuth. In 1953, the first machining tools for ultrasonic machining (USM) were produced which were used primarily for drilling and milling. Rotary ultrasonic machining (RUM) was introduced in 1964 by Percy Legge. In this case, the diamond abrasive was applied directly on the tool. At first, a workpiece performed the rotational motion, while the tool oscillated by ultrasonic frequency. This concept allows manufacturing only circular holes, and only for relatively small workpieces. However, development in this area led to the creation of machine tools, which allowed rotational movement of the tool during its current ultrasonic vibrations thanks to the rotational ultrasonic converter. It provides the possibility to machine also non-rotational shapes, and it is achieved with high precision. Scope of works has been expanded from primary drilling to milling of grooves, manufacturing of threads, as well as internal and external grinding [5].

The expansion in the usage of ultrasound in the field of machining occurred at the beginning of the 80s of the twentieth century, when the expansion of industrial applications of ceramics and composite materials occurred. Ultrasound also started to be used for thermally non-conductive alloys, non-metallic and electrically non-conductive materials, semiconductor materials and porous materials [11].

Ultrasonic machining can be divided into two fundamentally different groups. The first group consists of conventional ultrasonic machining, which uses a relatively soft non-rotating tool (e.g. stainless steel, bronze) and abrasive suspension with fine abrasives (e.g. diamond, CBN, Al_2O_3, SiC). This method is in the world known as ultrasonic machining (USM) [14].

The second group consists of rotary ultrasonic machining, which usually applies a synthetic diamond on the active surface of the rotating tool. Process fluid contains no abrasives and is used mainly for cleaning and cooling. This method is in the world known as rotary ultrasonic machining (RUM) [2]. Rotary ultrasonic machining achieved in comparison to conventional ultrasonic machining higher material removal rate, higher accuracy and lower parameters of roughness. This method is considered as a hybrid method that combines ultrasonic machining and grinding with diamond tools, which helps to achieve significantly better results than either method alone [10].

References

1. Astashev VK, Babitsky VI (2007) Ultrasonic processes and machines. Springer-Verlag, London
2. Ceramic industry (2012) Machining ceramics with rotary ultrasonic machining. http://www.cer amicindustry.com/articles/machining-ceramics-with-rotary-ultrasonic-machining. Accessed 9 Oct 2012
3. Galton (2013) Francis Galton. http://en.wikipedia.org/wiki/Francis_Galton. Accessed 14 Aug 2013
4. Hagen E (2009) Echolocation. ASU—Arizona State University—Ask a biologist. https://ask abiologist.asu.edu/echolocation. Accessed 20 Nov 2018
5. Khoo CY, Hamzah E, Sudin I (2008) A review on the rotary ultrasonic machining. In *Jurnal Mekanikal*
6. Koenig (2013) Rudolph Koenig. http://en.wikipedia.org/wiki/Rudolph_Koenig. Accessed 13 Aug 2013
7. Lengevin (2013) Paul Lengevin. http://en.wikipedia.org/wiki/Paul_Langevin. Accessed 15 Aug 2013
8. Lewis (2013) Lewis Fry Richardson. http://en.wikipedia.org/wiki/Lewis_Fry_Richardson. Accessed 16 Aug 2013
9. Loomis (2012) Alfred Lee Loomis. http://en.wikipedia.org/wiki/Alfred_Lee_Loomis. Accessed 19 Sept 2012
10. Maňková I (2000) Progresívne technológie (Publishers Vienela, Košice, 2000), pp 31–48. ISBN 80-7099-430-4
11. Mičietová A, Manková I, Velíšek K (2007) Fyzikálne technológie obrábania. Top trendy v obrábaní 2007(5):22–48
12. Spallazani (2013) Lazzaro Spallazani. http://en.wikipedia.org/wiki/Lazzaro_Spallanzani. Accessed 12 Aug 2013
13. Szabo TL (2004) Diagnostic ultrasound imaging: inside out. Elsevier, San Diego
14. USM (2012) Ultrasonic machining. http://www.ceramicindustry.com/articles/ultrasonic-mac hining. Accessed 3 Oct 2012
15. Wood (2012) Robert W. Wood. http://en.wikipedia.org/wiki/Robert_W._Wood. Accessed 19 Sept 2012

Chapter 3
The Current State in the Rotary Ultrasonic Machining

Abstract When ultrasound began to be used in industrial application, plenty of regularities and influences had to be discovered. The most of the researches in ultrasonic machining were focused on the evaluation of cutting forces, material removal rate, tool wear, surface roughness, accuracy, cutting temperature and power consumption during machining of hard and brittle materials and composite materials. There were made simulations of the process and analysed the removal process. The hybrid processes were developed with a purpose to improve the cutting process of alloys. Ultrasonic assistance of classic machining process was used to improve observable characteristics of machined alloys. Generally, it can be concluded that ultrasound positively affects every aspect of the machining process.

At present, the use of advanced materials is still rising. Between those materials is included for example technical ceramic because of its high hardness, high strength-to-weight ratio, chemical inertness, thermal stability, long life, etc. It is used primarily in the manufacture of semiconductors, in micro-electromechanical systems, health care, defence, aviation and electrical engineering. Components in these fields require high-dimensional accuracy, low roughness parameters, low tolerance, minimal thermal, chemical and mechanical influence of the surface and complex shape. Advanced materials are utilized especially in aircraft and spacecraft. Because of their mechanical or physical properties, they are usually hard-machinable. For applications like this have been developed advanced machining methods. These methods do not use standard cutting tools, but they use alternative energy sources, such as thermal, electrical, chemical or mechanical energy, or their combination (see Table 3.1) [10, 26].

One of the advanced machining methods is ultrasonic machining. It is based on mechanical–acoustic material removal. Ultrasonic machining was created as an advanced method of processing hard-machinable materials. Brittle and hard materials (such as ceramics, glass), which are problematically processed by conventional machining methods, are ideal for ultrasonic machining. It can machine electrically conductive materials as well as non-conductive materials [8, 26, 30].

The ultrasonic machining is characterized by the following parameters: oscillation frequency is in the range from 15 to 40 kHz. Power of supply is in the range 50 to

© The Author(s), under exclusive license to Springer Nature Switzerland AG 2021
M. Kuruc, *Rotary Ultrasonic Machining*, Manufacturing and Surface Engineering,
https://doi.org/10.1007/978-3-030-67944-6_3

Table 3.1 Division of advanced machining method based on their energy source [10]

Energy source		
Mechanical processes	Chemical or electrochemical processes	Thermal or electrothermal processes
Ultrasonic machining, abrasive jet machining, water jet machining, abrasive water jet machining	Chemical machining, photochemical machining, electrochemical machining, electrochemical grinding	Electrodischarge machining, laser beam machining, electron beam machining, ion beam machining, plasma arc machining

2400 W. The amplitude of oscillation is in the range from 13 to 100 μm. Size of abrasives is in the range from 9 to 50 μm. Abrasives are usually made of hard materials, such as aluminium oxide, silicon carbide, boron carbide, cubic boron nitride and synthetic diamond. The tool is made of softer materials such as stainless steel, molybdenum, Monel and brass. Water, kerosene, alcohol and mechanical oil are frequently used as the slurry. The slurry typically includes abrasives [10, 23].

Issue of ultrasonic machining is being solved by the industry as well as universities, because this machining method has great potential. Enforcement of ultrasonic machining is especially in the aerospace industry for machining advanced materials, in the medicine for creating dental replacements, in the watch industry for making fine mechanism, in the foundry for processing ceramic moulds, in automotive for cutting composite materials, etc. It is hard to imagine, for example, to create a thread into ceramics without rotary ultrasonic machining. Due to the multi-axis conception of machine tool, this technology can machine almost any material and almost any shape, while the high quality of the machined surface and low cutting force are present. Therefore, several companies start utilizing this method for the processing of advanced materials. It was accompanied by foundations of new corporations focused on manufacturing of the proper machine tools and cutting tools [9].

3.1 Relevant Industrial Corporations

Company *Ceramic Industry* uses rotary ultrasonic machining (RUM) for fast and high-quality machining of ceramics and glass. It is used by drilling hundreds of small holes (diameter 0.5 mm × 10 mm) in semiconductor materials such as silicon, quartz, sapphire and corundum (alumina—Al_2O_3). It finds application in the production of rods for laser equipment and intermediate products for optical fibres of quartz, glass, sapphire, ruby, aluminium nitride, silicon carbide, aluminium oxide and other materials. It also examines and seeks the most suitable materials for ultrasonic machining in terms of their prices and hard-machinability [2].

Sonic-Mill Company applies ultrasonic machining for drilling deep holes in the boride–silica glass, to produce thousands of holes in silicon, milling grooves in silicon carbide, for drilling holes close together and close to the edges to sapphire, and so on [22].

Company **Silfex** is considered to be the leader in ultrasonic machining because of their long experience and large amounts of the most advanced tools for ultrasonic machining. They can create accurate cavities of practically any shape in hard and brittle materials. In addition, the piece is not affected either thermally, chemically or electrically and therefore the final metallurgical, chemical and physical properties of the completed components remain unchanged. The company says it can produce more than 10,000 holes at the same time, with a diameter of 0.5 mm, and with different shapes in hard and brittle materials such as silicon, silicon carbide, silicon nitride, quartz, glass, corundum, sapphire and graphite. They can produce according to the special requirements of customers [20].

Among the well-known manufacturers of machines for ultrasonic machining (and also laser beam machining, five-axes machining, etc.) is company **DMG MORI**, which for this purpose has offered machine tools from Ultrasonic series [5], such as Ultrasonic 10 (Fig. 3.1a) and Ultrasonic 20 linear (Fig. 3.1b). Ultrasonic 10 is mainly used in dental medicine in the manufacture of dental prostheses. It is a 5-axis milling centre, whose spindle rotation frequency reaches 42,000 rpm. Ultrasonic 20 linear is used for precision and micromachining and can operate in 5 axes. Thanks to the spindle rotation frequency 60,000 rpm it is also used for high-speed machining [27–29].

For this kind of machine tools are required proper cutting tools. Company **SCHOTT Diamantwerkzeuge** is offering galvanic tools for hard and brittle material processing; diamond hollow drills and diamond cutters suitable for employment

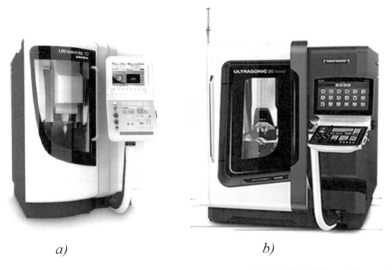

a) *b)*

Fig. 3.1 Ultrasonic milling machines made by company DMG MORI [5]. **a** Ultrasonic 10, **b** Ultrasonic 20 linear

Fig. 3.2   Cutting tools for rotary ultrasonic machining produced by Schott

under ultrasonic support; special tools to meet the highest standards of quality; tool development and ideal solutions; microdrills with external diameters from 0.3 mm onwards, with internal cooling. Examples of ultrasonic tools produced by Schott company are shown in Fig. 3.2. Those tools can be used to machine materials, such as quartz glass, float glass, optical glasses, ceramic material, boron carbide, carbon, dental ceramic, silicon, silicon carbides, titanium, steels, ferrite, hard metals, compound materials, zircon oxide, etc. [19].

Another producer of the cutting tools for rotary ultrasonic machining is *EFFGEN Schleiftechnik Lapport*. They produce different milling and drilling diamond tools for the manufacturing of glass and ceramic, as well as grinding tools and wheels, polishing tools, saw blades, dressing tools and rolls, etc. Those tools are offered in different sizes and shapes [6].

3.2   Researches in the Field of Ultrasonic Machining

Research in the field of rotary ultrasonic machining (RUM) is in a relatively early stage, but there are lots of interesting discoveries so far. Benefits of this method predetermine it to processing of materials, which are difficult-to-cut, where a high quality of the surface is required. Therefore, it starts to be attractive for advanced industries. Researchers at companies and universities are developing and improving this method, looking for additional applications and expanding of knowledge the process itself.

For example, *Pei* **et al**. investigated cutting forces for RUM of brittle materials. They created a mechanistic model and compare results from the simulation with experimental verification. They have found out that trends of predicted influences of input variables on cutting forces obtained by their model correspond well with the trends obtained experimentally: cutting forces are decreasing with increasing spindle speed, amplitude of vibration and size of abrasive particles; and it is increasing with the increment in feed rate and concentration of abrasive particles. But the highest influence on cutting forces has the abovementioned spindle speed and feed rate, as shown in Fig. 3.3. Vibration amplitude 25 µm, vibration frequency 20 kHz, abrasive size 0.125 mm, abrasive concentration 100, diamond abrasive, metal type of bond, and alumina workpiece have been used in the experiment. For the left figure (Fig. 3.3a), 6 mm min^{-1} feed rate has been used, and for the right figure (Fig. 3.3b), 2000 rpm spindle speed has been used. These results should serve to predict cutting temperatures, tool wear and value of roughness of machined surface [18].

Pei in cooperation with *Ferreira* has also experimentally investigated rotary ultrasonic milling. Usually, RUM is used for machining circular holes. Therefore, they want to expand knowledge about usage of RUM for other application, which is utilized in industry for the manufacturing of advanced ceramics. Many modern engineering sectors are using ceramics for critical application. Therefore, research in this area is needed. They have found out that it is only the depth of cut and feed rate that have an influence on the material removal rate of ultrasonic face milling. During RUM, two material removal modes are present: ductile mode and brittle fracture. The higher the percentage of ductile mode is, the lower the value of surface roughness is obtained [13].

Ya **et al**. deal with the analysis of the rotary ultrasonic machining mechanism. They consider the impact and grinding of the abrasive in the tool tip on the machined surface as the main factors in the material removal rate (MRR) for ultrasonic machining.

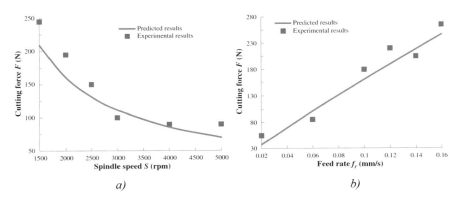

Fig. 3.3 Influence of spindle speed and feed rate on cutting force [18]. **a** Spindle speed—cutting force influence, **b** feed rate—cutting force influence

They find three material removal mechanisms: impacting, abrasion and ultrasonic cavitation. They have found out that the biggest impact on MRR has a static load, the grid of the abrasive, concentration of the abrasive, mechanical properties of a workpiece, material of a tool, spindle speed and feed rate. They also simplified the equation of calculation of material removal rate on [31]:

$$\text{MRR} = f \cdot \rho \cdot V \tag{3.1}$$

where

MRR   is material removal rate [kg s^{-1}],
f is ultrasonic vibration [s^{-1}],
ρ is density of workpiece [kg m^{-3}],
V is total volume of removed material, removed in each vibration cycle [m^3].

Thoe, Aspinwall and Killey combine ultrasonic machining (USM) process with EDM (electrical discharge machining) process. They were dealing with the creation of small holes (up to 1 mm) into nickel alloy coated by non-conductive ceramics. This material is used in the aerospace industry, especially for gas turbine components. Because of electrically nonconductive coating, it is not suitable to use EDM alone, and laser beam machining (LBM) could cause the formation of microcracks and thermal stresses in the workpiece. Therefore, USM looks like as an ideal solution. They compared MRR of EDM process of creating holes with and without the assistance of ultrasound. The results are shown in Fig. 3.4. The main used parameters were: ultrasonic amplitude 10 μm, ultrasonic frequency 23 kHz and electrical current

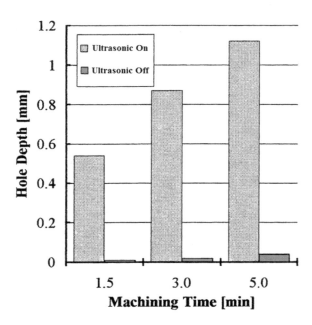

Fig. 3.4 Effect of ultrasonic assistance on MRR on coated Ni alloy [25]

5 A. Tool material for EDM is not proper for USM and vice versa. The best parameters have been achieved, when mild steel as a tool material has been used. During ultrasonic-assisted EDM has been obtained a better surface quality of the holes in comparison with standard EDM process. Moreover, ultrasound helps to stabilize the EDM process. However, tool life has been decreased down to 20–25 min [25].

Zeng, Li, Pei and Treadwell observed tool wear during rotary ultrasonic machining (RUM) of advanced ceramics. RUM is a cost-effective machining method for advanced ceramics, where tool wear was not thoroughly investigated. Therefore, the tool wear on the end face and lateral face, as well as cutting forces, has been examined. Silicon carbide (SiC) has been used as a workpiece. They found out that the end face of the tool is much more severe than that one on the lateral face. Also, friction wear and bond fracture have been observed, while grain fracture (which is common for grinding) has not been observed. Tool wear has two stages: at first, friction wear is dominant, characterized by increasing of cutting force; and then the bond fracture is dominant, characterized by a decrease in cutting force [33].

Yu, Hu and Rajurkar investigated material removal and surface roughness in micro-ultrasonic machining of silicon. Experimental results show that the machining speed is decreasing after the static load increases beyond a certain value. They have found out it is caused by debris when it cannot be removed fast enough from the working area. This debris also causes increasing of roughness, when the very small abrasive is used. Also, the rotation of a tool can improve the process. When the tool is not rotating, there could be locally concentrated abrasive particles, wherein some parts the surface roughness Ra equals 0.075 μm, while in some other parts roughness Ra equals 0.533 μm. During experiments, the tungsten tool $\varnothing$ 0.095 mm has been used with frequency 39.5 kHz and amplitude 1 to 1.5 μm. Rotation of tool (when has been used) was 3000 rpm. Abrasive was made of polycrystalline diamond with size 0.25 to 3 μm carried in water. The workpiece has been made of silicon [32].

Nath, Lim and Zheng deal with the influence of material removal mechanisms on the hole integrity in ultrasonic machining (USM) of structural ceramics. Elementary material removal mechanism for USM is microchipping via microcracks. It is caused by rapid mechanical indentations by abrasive grids. It consists of hammering action, impact action and cavitation–erosion action (depends on material properties). They mainly investigated the influence of these fundamental mechanisms on the hole integrity, such as entrance chipping, surface roughness and subsurface damage, for three kinds of workpieces: silicon carbide SiC, zirconia ZrO_2 and alumina Al_2O_3. Figure 3.5 shows the influence of material and diameter of the hole on entrance chipping. The largest chipping occurs in alumina due to its low fracture toughness and flexural strength. Created cracks are about 2 to 4 times bigger than the radius of used abrasives. The portion of cracks appears as surface roughness and subsurface damage. They cannot be eliminated, but they can be minimized by smaller abrasive grids, because the length of cracks depends on grid size. However, the decreasing of grid size also causes a decrease in productivity [11].

Gong, Fang and Hu investigated tool life in rotary ultrasonic side milling of hard and brittle materials. They have found out that rotary ultrasonic machining (RUM) causes lower tool wear than grinding under the same conditions. In addition, RUM

SiC

ZrO$_2$

Al$_2$O$_3$

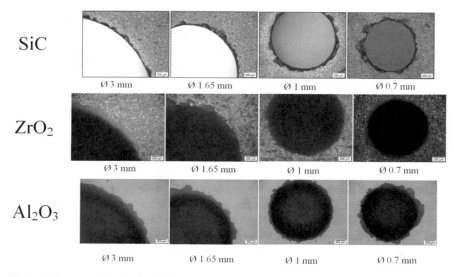

Fig. 3.5 Entrance chipping for different materials and different diameter of hole [11]

is much faster than grinding. After RUM, there has not been any tool wear observed. Overleaf, after grinding, significant tool wear has been observed. It is explained by ultrasonic vibration. During RUM is grid rotating and simultaneously moving up and down at ultrasonic sinusoid motion, as shown in Fig. 3.6. Therefore, it is moving faster than the grid in grinding. It also causes that more overlaps can be found in RUM, which means grid would cut less material in comparison to grinding, at the

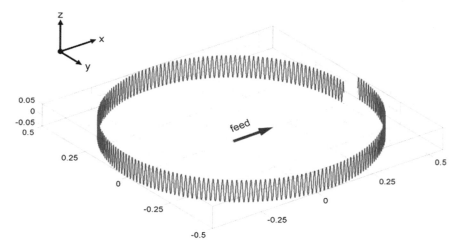

Fig. 3.6 Trajectory of diamond abrasive for RUM using side milling [7]

same material removal rate (MRR). Therefore, it is considered as the main reason for the decrease in tool wear [7].

Pei, Hu, Zhang and Treadwell also investigated the relationship between material removal rate (MRR) and controllable machining parameters in rotary ultrasonic machining (RUM). They have found out that static force, vibration amplitude and grid size have the most significant effect on the MRR. The rotation speed of the tool and the number of abrasive particles have no significant influence on the MRR. MRR increases with increasing of static force, increasing vibration amplitude and increasing of grid size [15].

Pei and Ferreira were modelling ductile mode of material removal in rotary ultrasonic machining (RUM). In RUM of hard and brittle materials, such as ceramics, there exist two material removal models—brittle fracture mode and ductile mode. Brittle fracture mode is a dominant one. Someone would expect only brittle fracture mode during machining of brittle materials; however, a ductile mode is also present. It is a minor mode, but it also influences the machining process. Due to ductile material removal mode, "crack-free" RUM of ceramics is possible. Therefore, they decided to investigate ductile mode for RUM of magnesia-stabilized zirconia. They created a mathematical model and compared it with experiments. They found out that increment in amplitude increases material removal rate (MRR) as well. However, increasing amplitude after a certain value will not effectively increase MRR. Increasing of static force increases MRR; however, too large static force will change the material removal from the ductile mode into the brittle fracture mode (mathematical model of ductile mode can be applied only during machining where ductile material removal mode is present). Increasing the rotational speed of the tool increases MRR, but the effect of increase depends on the static force—large static force causes a stronger effect of rotational speed. Decreasing the number of abrasive particles will increase MRR. However, too low number of abrasive particles will change material removal from ductile mode into brittle fracture mode. Decreasing the size of abrasive particles increases MRR, but also, too small size of abrasive particles will change material removal from ductile mode into brittle fracture mode. These mathematical dependences have been verified by experiments. Following parameters have been used: magnesia-stabilized zirconia workpiece, water coolant, spindle speed 3000 rpm, vibration frequency 20 kHz and vibration amplitude 23 µm. Calculated MRR (by mathematical model) agrees well with experimental MRR, especially in low force range. In high static force range, the error prediction increases. It is caused by predominating of material removal mode from ductile mode to brittle fracture mode. A ductile mode is dominant when the static force is lower than 200 N [12].

Aspinwall, Thoe and Wise have made a review on ultrasonic machining. Variations on ultrasonic machining basic configuration include: rotary ultrasonic machining; ultrasonic machining combined with electrical discharge machining; ultrasonic-assisted conventional or non-conventional machining; and non-machining ultrasonic applications such as cleaning, welding, forming, coating, chemical processing, etc. They have found out that RUM enhances MRR and workpiece accuracy, reduces cutting forces and increases tool life. MRR in RUM is 6–10 times higher than in diamond grinding under similar conditions, and up to 4 times higher than in standard USM, as shown in Fig. 3.7. In this figure, MRR of USM and RUM for different material are compared. These materials differ by hardness-to-Young's modulus ratio (H/E) [1].

In addition, they investigate the accuracy and roughness of the machined surface. They have found out that the surface roughness is being improved by the increase of static load. Differences between USM and RUM are shown in Fig. 3.8. There are compared accuracy (out-of-roundness) and surface roughness achieved by these methods for different materials. Lower value stands for a better result [1].

Pei, Ferreira, Kapoor and Haselkorn investigate the rotary ultrasonic machining (RUM) for face milling of ceramics. For some high precision components made of advanced ceramics, the cost of machining can constitute 90% of the total cost. Therefore, for machining of ceramics, the RUM has high potential. RUM can achieve higher MRR and similar accuracy as the diamond grinding. RUM is usually used for

Fig. 3.7 Effect of H/E on MRR for USM and RUM [1]

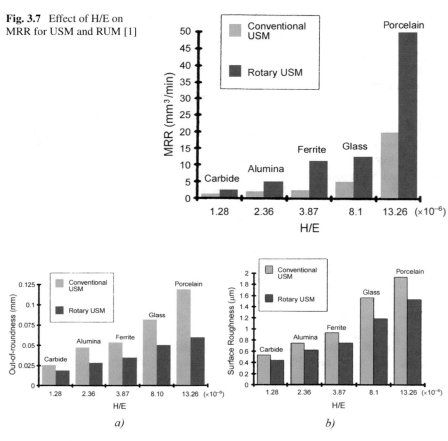

Fig. 3.8 Effect of hardness-to-Young's modulus ratio on accuracy and roughness [1]. **a** Effect of H/E on accuracy, **b** effect of H/E on surface roughness

manufacturing of circular holes or cavities, but it can machine flat surfaces or milling slots. However, these technological operations change the material removal mechanism. Main material removal mechanisms for ultrasonic drilling are hammering, abrasion and extraction. But ultrasonic milling could reduce the number of material removal mechanisms only to one—abrasion. It has resulted in a reduction of MRR, and the process is more similar to conventional diamond grinding. However, Pei et al. found a solution—if the bottom of a tool is not cylindrically shaped, but it is of conical shape, all three material removal mechanisms should be kept. But, if the angle between the machined surface and the conic surface is too high, wearing of tool increases. If this angle is too small, the maximum depth of the cut is reduced. This angle has usually value about 15° [14].

Cong, Pei, Deines, Srivastava, Riley and Treadwell also study rotary ultrasonic machining of carbon fibre-reinforced plastic (CFRP) composites. These kinds of composites are utilized in the aerospace industry, and it is very difficult to machine them. Fortunately, RUM can machine them both successfully and cost-effectively. There has already been compared RUM and twist drilling for drilling CFRP composites, and it was found out that RUM reduces the cutting force, torque, cutting temperature, workpiece delamination and tool wear. In that experiment, RUM is using cold air as coolant (not liquid such as water, as usually). They have found out that the highest power consumption has a coolant pump, and it does not depend on ultrasonic power (about 70%). Air compressor has also constant energy consumption (about 11%). The power consumption of ultrasonic power supply is increasing with the increase of ultrasonic power from 0% (at 0% of ultrasonic power) up to 16% (at 80% of US power). Power consumption of spindle motor is decreasing from 20% (at 0% of US power) to 3% (at 80% of US power). Power consumption of the whole ultrasonic system is almost constant, too (about 1 W h). Also, cutting force decreases to about 20% and torque decreases to about 40%. This dependence is recorded in graphs in Fig. 3.9. Following parameters have been used: tool rotation speed 3000 rpm and feed rate 0.5 mm/s [3].

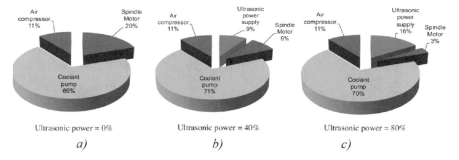

Fig. 3.9 Power consumption percentage of each component of ultrasonic system under different settings of ultrasonic power [3]. **a** without ultrasound, **b** 40% of ultrasonic power, **c** 80% of ultrasonic power

Change of tool rotation speed also affects power consumption. Power consumption of coolant pump, air compressor and ultrasonic power supply are almost constant, but the power consumption of spindle motor increases from 1% (at 1000 rpm) up to 15% (at 5000 rpm). Next parameter influencing power consumption is the feed rate. Power consumption of air compressor is independent of the feed rate. Power consumption of other parameters is dramatically decreasing with the increasing feed rate. In this case, power consumption of whole ultrasonic system is decreasing from value 4.8 W h (at 6 mm/min) to 0.8 W h (at 42 mm/min) [3].

Besides other things, *Pei, Li, Jiao, Deines and Treadwell* study rotary ultrasonic machining of ceramic matrix composites (CMC). These materials have high-temperature stability, high thermal shock resistance and low mass. Therefore, they are utilized in applications in aerospace, power generation, ground transport, nuclear, environmental and chemical industries; however, they are hard-machinable. Pei et al. compared RUM and diamond drilling of CMC. They investigate MRR and cutting forces. Water jet machining (WJM) of this kind of material often produces delamination, and laser beam machining (LBM) produces thermal stress and heat-affected zones. In experiments, three kinds of materials have been used: silicon carbide (SiC) fibre with converted phenolic resin char, densified by carbon chemical vapour infiltration, this carbon matrix contains boron carbide (BC) filler (in this issue labelled as CMC #1); SiC fibre with SiC partial matrix, which was further densified by the melt-infiltration process with Si metal (in this issue labelled as CMC #2); and sintered alumina (Al_2O_3). As shown in Fig. 3.10, RUM provides much lower cutting forces than diamond drilling. Moreover, about 10% higher MRR has been achieved during

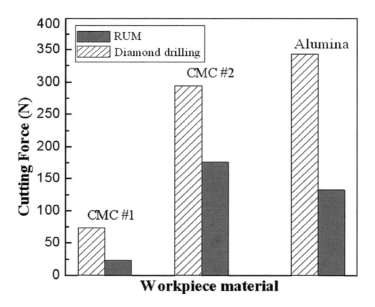

Fig. 3.10 Comparison of cutting forces for RUM and diamond drilling [16]

RUM. However, during RUM the edge-chipping can occur on the entrance and/or exit of holes. But it can be reduced or even prevented by suitable adjusting of machining parameters (higher spindle speed, lower feed rate) or by using sharp tools [16].

Pei with his scientific staff (*Wang, Cong, Gao and Kang*) also investigated RUM of potassium dihydrogen phosphate (KDP) crystal. KDP crystal is utilized especially for important electro-optic parts. High quality of the machined surface is usually required for this kind of parts. However, KDP crystal is hard-machinable, because it is soft, brittle and anisotropic. In addition, KDP dissolves in water and therefore spindle oil as a coolant must be used. They found out that surface roughness is increasing with increasing spindle speed and feed rate. Ultrasonic power has no significant influence on roughness. It is different from the results reported in the case of other materials. Increasing of roughness by the increase of spindle speed can be explained by used milling strategy—plunging. Increasing of spindle speed might cause a vibration of spindle, and this increases roughness. There have also been investigated the influence of the shape of a tool (see Fig. 3.11) on the surface roughness. They found out the smoothest surface has been achieved by the tool with a chamfered corner. When tool with two slots has been used, better surface parameters have been reached at a higher feed rate in comparison with a tool with common shape. Size of diamond grains has not as significant effect on surface integrity as a tool design itself [17].

Curodeau, Guay, Rodrigue, Brault, Gagné and Beaudoin investigated ultrasonic abrasive micromachining with thermoplastic tooling. Application of this method is mainly to remove heat-affected zone, remove surface waviness features and decrease surface roughness. They achieve surface roughness Ra 1.28 μm during machining (hammering mode) mould cavity. While tool is made of thermoplastic polymer, when the tool is worn out, it can be replaced or reprocessed (only rework its surfaces). They found out that larger grain size makes the process more effective. However, when too large grain size is used, the tool must regularly be lifted in order to refresh the gap with

a)   *b)*

Fig. 3.11 Different tool design used in experiments [17]. **a** Common shape (left) and chambered corner (right), **b** with two slots (left) and common shape (right)

new abrasive particles and maintain the process effectiveness. Moreover, polymer tool does not develop burrs at large amplitudes, in contrast with copper or brass tools. However, burrs can occur on polymer tool at higher pressure. During ultrasonic polishing (non-contact mode) of wire EDM surface, Ra 0.15 μm can be achieved. That means the thermoplastic tool can be efficiently used in micromachining and micropolishing of tool steel surface [4].

Stoll and Neugebauer study ultrasonic application in drilling. Ultrasound has been applied in terms of drilling due to the problems, which occur during conventional drilling of deep holes and overlapped workpieces, such as chip formation, chip removal, burr formation, tool stress and process reliability. Some of the advantages of ultrasound support are decrement in cutting forces and chip braking. During conventional drilling, the cutting speed at the tool centre is equal to zero and cutting conditions are accordingly bad; however, when drilling is supported by the ultrasound, cutting speed at the centre of the tool is different from zero, although limited in time. This fact enhances the whole machining process. Ultrasonic vibration causes changes in working angles of the tool. It improves chip formation and reduces friction on the tool, as well as cutting momentum in the process. Therefore, researchers found out that cutting force and momentum could be reduced by 30 to 50% when the ultrasound supports drilling process of silumin (Al–Si alloy). This fact enables the increase in tool life [24].

Singh and Khamba investigated ultrasonic machining of titanium and its alloys. Titanium is lightweight and high-strength metal. It has the highest strength-to-weight ratio of all metals. Titanium has good corrosion resistance, high toughness (even at low temperatures), high melting point (1668 °C) and poor thermal conductivity. Due to its properties, titanium has high utility in the manufacturing sector. However, these properties make titanium hard-machinable. Therefore, it is usually machined by the non-conventional processes, such as electric discharge machining (EDM). EDM gives good MRR; however, accuracy and surface finishing are not very good. Accuracy, as well as machining efficiency, could be improved, when EDM process is combined with USM. USM itself can decrease cutting forces and temperatures and therefore improve the machining process. At lower temperature, better surface finishing is achieved. Therefore, the selection of operating parameter levels is critical in order to achieve acceptable productivity. They investigated surface roughness, MRR and tool wear rate during USM of pure titanium (Grade 2) and titanium alloy Ti-6Al-4 V (Grade 5) for different abrasive materials. Results are shown in Table 3.2 [21].

According to presented researches, there can be seen that most of them were focused on the evaluation of cutting forces, material removal rate, tool wear, surface roughness, accuracy, cutting temperature and power consumption during machining of hard and brittle materials (Al_2O_3, SiC, Si, ZrO_2, SiO_2, KDP, advanced ceramics) and composites (CFRP, CMC). There were made simulations of the process and analysed the removal process. The controllable machining parameters, such as cutting speed (spindle speed), vibration amplitude, abrasive size, feed rate, abrasive concentration, depth of cut and static load (force) were changed, as well as tool shape and material. Hybrid processes (USM + EDM) were developed as well with a purpose to

Table 3.2 Data from titanium alloys ultrasonically machined using Ø5 mm solid tool [21]

Workpiece material	Tool material	Recommended abrasive	Surface roughness, Ra (μm)	Material removal rate (mg/min)	Tool wear rate (mg/min)
Ti Gr. 2	Stainless steel	Al_2O_3	0.48	5.00	10.10
		SiC	0.31	4.13	9.20
		B_4C	0.46	2.63	7.13
Ti Gr. 5	Stainless steel	Al_2O_3	0.44	3.71	8.38
		SiC	0.46	2.77	5.55
		B_4C	0.56	2.47	6.63

improve the cutting process of alloys (Ni, Ti), or just ultrasonic assistance of classic machining process was used to improve observed characteristics of the machined alloy (Al–Si). Generally, it can be concluded that ultrasound positively affects every aspect of the machining process.

References

1. Aspinwall DK, Thoe TB, Wise MLH (1998) Review on ultrasonic machining. Int J Mach Tools Manuf 38(4):239–255
2. Ceramic industry (2012) Machining ceramics with rotary ultrasonic machining. [cit. 02. 10. 2012]. Available at: http://www.ceramicindustry.com/articles/machining-ceramics-with-rotary-ultrasonic-machining. Accessed 2 Oct 2012
3. Cong WL, Pei ZJ, Deines TW, Srivastava A, Riley L, Treadwell C (2012) Rotary ultrasonic machining of CFRP composites: a study on power consumption. In: Ultrasonics. http://dx.doi.org/10.1016/j.ultras.2012.08.007
4. Curodeau A, Guay J, Rodrigue D, Brault L, Gagné D, Beaudoin L-P (2008) Ultrasonic abrasive μ-machining with thermoplastic tooling. Int J Mach Tools Manuf 48:1553–1561
5. DMG MORI (2012) https://en.dmgmori.com/. Accessed 5 Oct 2012
6. Effgen (2018) Lapport Schleiftechnik. http://effgen.eu/de/effgen-schleiftechnik. Accessed 30 Nov 2018
7. Gong H, Fang FZ, Hu XT (2010) Kinematic view of tool life in rotary ultrasonic side milling of hard and brittle materials. Int J Mach Tools Manuf 50:303–307
8. Humár A (2005) Technologie I – Technologie obrábění. 3. Část. Brno
9. KSU (2013) Kansas State University. http://www.k-state.edu/. Accessed 19 Aug 2013
10. Maňková I (2000) Progresívne technológie (Publishers Vienela, Košice), pp. 31–48. ISBN 80-7099-430-4
11. Nath Ch, Lim GC, Zheng HY (2012) Influence of the material removal mechanisms on hole integrity in ultrasonic machining of structural ceramics. Ultrasonics 52:605–613
12. Pei ZJ, Ferreira PM (1998) Modeling of ductile-mode material removal in rotary ultrasonic machining. Int J Mach Tools Manuf 38:1399–1418
13. Pei ZJ, Ferreira PM (1999) An experimental investigation of rotary ultrasonic face milling. Int J Mach Tools Manuf 39:1327–1344
14. Pei ZJ, Ferreira PM, Kapoor SG, Haselkorn M (1995) Rotary ultrasonic machining for face milling of ceramics. Int J Mach Tools Manuf 35(7):1033–1046
15. Pei ZJ, Hu P, Zhang JM, Treadwell C (2002) Modeling of material removal rate in rotary ultrasonic machining: designed experiments. J Mater Process Technol 129:339–344

16. Pei ZJ, Li ZC, Jiao Y, Deines TW, Treadwell C (2005) Rotary ultrasonic machining of ceramic matrix composites: feasibility study and deign experiments. Int J Mach Tools Manuf 45:1402–1411
17. Pei ZJ, Wang Q, Cong W, Gao H, Kang R (2009) Rotary ultrasonic machining of potassium dihydrogen phosphate (KDP) crystal: an experimental investigation on surface roughness. J Manuf Process 11:66–73
18. Pei ZJ et al (2012) A cutting force model for rotary ultrasonic machining of brittle materials. Int J Mach Tools Manuf 52:77–84
19. Schott (2013) Diamond tools: ultrasonic. http://www.schott-diamantwerkzeuge.com/ultrasonic.html. Accessed 1 Sept 2013
20. Silfex (2012) http://www.silfex.com/services_2_4.html. Accessed 19 Nov 2012
21. Singh R, Khamba JS (2006) Ultrasonic machining of titanium and its alloys: a review. J Mater Process Technol 173:125–135
22. Sonic-Mill (2012) Machining the unmachinable. http://www.sonicmill.com/contract.html. Accessed 3 Oct 2012
23. Spišák E (2011) Strojárske technológie. University of Technology, Košice. ISBN: 9788055308203
24. Stoll A, Neugebauer R (2004) Ultrasonic application in drilling. J Mater Process Technol 149:633–639
25. Thoe TB, Aspinwall DK, Killey N (1999) Combined ultrasonic and electrical discharge machining of ceramic coated nickel alloy. J Mater Process Technol 92–93:323–328
26. Top trendy v obrábaní V (2007). s. l.: Media/ST, 2007. ISBN: 8096895472
27. Ultrasonic (2012) Machining of advanced materials. http://us.dmg.com/us,ultrasonic,overview. Accessed 5 Oct 2012
28. Ultrasonic_10 (2012) The most compact DMG machine center. http://us.dmg.com/us,ultrasonic,ultrasonic10?opendocument. Accessed 5 Oct 2012
29. Ultrasonic_20 linear (2016) Optimum versatility with a combination of technology—quality without compromise. https://en.dmgmori.com/products/machines/advanced-technology/ultrasonic/ultrasonic-linear/ultrasonic-20-linear. Accessed 5 Oct 2016
30. USM (2012) Ultrasonic machining. http://www.ceramicindustry.com/articles/ultrasonic-machining. Accessed 3 Oct 2012
31. Ya G et al (2002) Analysis of the rotary ultrasonic machining mechanism. J Mater Process Technol 129:182–185
32. Yu Z, Hu X, Rajurkar KP (2006) Influence of Debris accumulation on material removal and surface roughness in micro ultrasonic machining of silicon. CIRP Ann Manuf Technol 55(1):201–204
33. Zeng WM, Li ZC, Pei ZJ, Treadwell C (2005) Experimental observation of tool wear in rotary ultrasonic machining of advanced ceramics. Int J Mach Tools Manuf 45:1468–1473

Chapter 4
Prognoses in the Field of Rotary Ultrasonic Machining

Abstract The research in ultrasonic machining is a never-ending process. New discoveries lead to new devices and applications. And they request additional research to fully understand their processes. And during the research could be discovered additional phenomena, which could find application in the future. Moreover, development in the analytical sector allows obtaining new or more precise analyses of the process. The following chapter tries to briefly introduce possible developments in the field of rotary ultrasonic machining.

Nowadays, rotary ultrasonic machining is mainly used for making round holes, but research is starting to focus on the planar machining, where the tool is machining not only vertically, but also horizontally. In this case, in terms of material removal rate, it is more suitable, when the tool has a conical shape instead of the classical cylindrical one [1, 4].

Other development is focused on the study of five-axis ultrasonic milling of free form surfaces (FFS), where the tool does not operate only perpendicularly considering to surface of a workpiece, as is common during three-axis ultrasonic milling. It should affect the spreading of ultrasonic waves, their reflection, as well as absorption.

Further scientists based their research on high-speed ultrasonic machining, where the cutting speed of tool can reach several hundred, up to several thousand meters per minute. It is commonly known that milling assisted by ultrasound achieves better results in comparison with traditional milling, as well as high-speed milling also achieves better results than traditional milling. Therefore, they expect another improvement when they merge these two advanced technologies.

In terms of increasing of productivity, there performed the experiments of rotary ultrasonic machining with high depth of cut (DOC). Simple increasing of DOC would cause increasing of cutting force; therefore, there has to be feed rate compensation. However, even with reduced feed rate, significant material removal rate should be observed.

Some researchers investigate micro-ultrasonic machining, because of increasing requirements to manufacture very small features with high quality of surface into advanced ceramics and similar materials (especially in the electronics industry) [6].

© The Author(s), under exclusive license to Springer Nature Switzerland AG 2021
M. Kuruc, *Rotary Ultrasonic Machining*, Manufacturing and Surface Engineering, https://doi.org/10.1007/978-3-030-67944-6_4

There is also the research of ultrasonic machining for machining of advanced materials, such as composites, cubic boron nitride and so on, besides machining of traditional materials for ultrasonic milling, such as ceramics and glass.

Some scientists are trying to use cold air or oil mist as cooling medium instead of standard liquid (water). This change should reduce the costs associated with cutting fluids [2].

Another research area is focused on the fixation of workpieces. Classic clamping is being replaced by fixation with wax. It is possible due to the low cutting force of the rotary ultrasonic machining process. Therefore, wax can create enough force to fix the workpiece to the machine tool. Other clamping systems can use a vacuum to fix the workpiece.

An important part of RUM system is the tool. Therefore, some experiments based on the tool design have been implemented. There has been investigated the influence of the chamfered corner as well as slots on the active part of the tool [5].

New strategies of ultrasonic machining are also a very interesting area of research. Machining process during ultrasonic machining has not the same behaviour as the conventional machining process. Therefore, for improving the process itself (increasing tool life, reduction of machining time, decreasing surface roughness, improving accuracy) it is necessary to use proper machining parameters and strategies [7].

For rotary ultrasonic machining, hard and brittle materials are suitable. Therefore, there are some experiments with liquid nitrogen as a coolant. Cryogenic conditions make machined material even harder and more brittle, and therefore, it could be even more suitable for ultrasonic machining. Moreover, materials which are not so suitable for RUM could become to be.

Developing of artificial intelligence (and Industry 4.0) brings many benefits. The first experiments of its implementation into the machining process, even in the ultrasonic machining, begun. This system allows to predict occasions and reduce/avoid process errors. It increases the productivity and quality of the process [3].

Another possibility of development in the ultrasonic machining could be implementation other ultrasonic application into one machine—achieving universal machine available to machining, welding and cleaning. This would require an ultrasonic generator with a higher range of frequencies and additional devices to transform ultrasonic energy into the form suitable for each ultrasonic application separately.

References

1. Ceramic industry (2012) Machining ceramics with rotary ultrasonic machining. http://www. ceramicindustry.com/articles/machining-ceramics-with-rotary-ultrasonic-machining. Accessed 2 Oct 2012
2. Cong WL, Pei ZJ, Deines TW, Srivastava A, Riley L, Treadwell C (2012) Rotary ultrasonic machining of CFRP composites: a study on power consumption. In: Ultrasonics. http://dx.doi. org/10.1016/j.ultras.2012.08.007
3. Mahamad AKB (2010) Diagnosis, classification and prognosis of rotating machine using artificial intelligence. Kumamoto University, Japan, p 114

4. Pei ZJ, Ferreira PM, Kapoor SG, Haselkorn M (1995) Rotary ultrasonic machining for face milling of ceramics. Int J Mach Tools Manuf 35(7):1033–1046
5. Pei ZJ, Wang Q, Cong W, Gao H, Kang R (2009) Rotary ultrasonic machining of potassium dihydrogen phosphate (KDP) crystal: an experimental investigation on surface roughness. J Manuf Process 11:66–73
6. Yu Z, Hu X, Rajurkar KP (2006) Influence of Debris accumulation on material removal and surface roughness in micro ultrasonic machining of silicon. CIRP Ann Manuf Technol 55(1):201–204
7. Zvončan M (2012) Research of edgechipping in rotary ultrasonic machining. Dissertation thesis, STU, MTF-10906-32462, 108 pp

Chapter 5
Theory of the Ultrasound

Abstract To fully understand the processes in ultrasonic machining, it is necessary to understand the theory of the ultrasound first. The material affects the spreading of the ultrasonic waves. The damping of the undulation affects the tool material selection. The modulus of elasticity affects the machinability of the workpiece material. To sufficient ultrasonic machining, the amplitude of the undulation has to be high enough. With those and additional knowledge is possible to construct devices for ultrasonic machining. For their proper operation, they have to consist of ultrasonic, transducer, concentrator and tool. Correct selection of the tool material, workpiece material, ultrasonic set-up and machining parameters cause benefits during the cutting process.

The sound is a mechanical undulation in the matter environment, which is characteristic by its specific frequency. The scientific discipline which is engaged in sound is called acoustics. The hearable sound occurs in the range from 20 Hz to 20,000 Hz (this range is changing with an age of the person). Sound which has a lower frequency than 20 Hz is called the infrasound. Over 20 kHz, the ultrasound is present. If the frequency of sound is higher than 10 THz (10^{10} kHz), we are talking about hypersound. The hypersound is utilized, for example, in molecular acoustics [4, 14, 17].

Ultrasound can be divided based on the intensity of ultrasound undulation on passive ultrasound (ultrasound with low amplitudes and high frequency) and active ultrasound (ultrasound with high amplitudes and high intensity). Passive ultrasound does not cause either physical or chemical changes in the environment where it is present. Therefore, it is utilized especially in measuring, inspecting and controlling technologies. Active ultrasound affects properties of the environment in which it is present; therefore, it is utilized in areas where the acceleration of particular processes or increase of quality of performed activity are needed, respectively in other areas, which utilize positive properties of ultrasound energy. Active ultrasound is used, for example, in ultrasonic machining, welding or cleaning [16, 17].

A fundamental property of ultrasonic undulation is harmonic oscillation. Therefore, immediate deflection of the affected particle could be calculated by the equation [17]:

© The Author(s), under exclusive license to Springer Nature Switzerland AG 2021
M. Kuruc, *Rotary Ultrasonic Machining*, Manufacturing and Surface Engineering,
https://doi.org/10.1007/978-3-030-67944-6_5

$$\xi = A \cdot \sin(\omega \cdot t + \varphi) \tag{5.1}$$

where

ξ is immediate deflection (m),
A is amplitude (the highest value of deflection) (m),
ω is angular velocity (rad s^{-1}),
φ is phase shift (rad), and
t is time (s).

5.1   Ultrasonic Undulation

Ultrasonic energy causes different effects in the environment, such as the pressure of ultrasonic radiation, intensive alternating pressure, cavitation, thermal impact, deformation, mechanical and electrochemical effects and so on, which can cause mechanical, physical and chemical changes of the environment, where ultrasound is spread. Spreading of ultrasound is based on transferring of energy between particles of the environment (atoms, molecules, volume elements), while particles themselves do not displace, but only oscillate around their equilibrium location. This process is also called undulation. Ultrasonic undulation is therefore based on elastic deformations of the environment, while it is transferred as a flow with specific space density. This process can be called the ultrasonic radiation, or gradual undulation. If this undulation stays in space of oscillating solid (ultrasonic emitter), we call this effect the stationary undulation, or quaver, or oscillation, or vibration [7, 17].

Spreading of ultrasonic waves has always the space character, while a considerable amount of mutual adjacent particles is oscillating in the same phase. These oscillating particles, which vibrate by the same phase, constitute the wave surface. Wave surfaces could be divided into the planar, cylindrical and spherical ones. The shape of wave surface depends on the shape of the ultrasonic source. If the ultrasonic source looks like the surface, resultant wave surface is planar in shape. If the ultrasonic emitter is straight, resultant wave surface is cylindrical. And if the ultrasonic emitter is a point in shape, resultant wave surface is spherical [7, 17].

When considering the properties of the environment, there could be spread either longitudinal or traverse waves. Longitudinal undulation is an oscillation of elements of the environment in the direction of spreading of the undulation. It is manifested by condensation and dilution of environmental particles (see Fig. 5.1a). This kind of undulation is the most frequently occurred one and to be formed, it demands the only environment that is large enough. Its velocity in solid matter is affected by tension modulus of elasticity and pressure modulus of elasticity. Traverse undulation is an oscillation of environment elements in the direction perpendicular to the direction of spreading of the undulation. It is manifested by the occurrence of positive and negative peaks in the horizontal plane (see Fig. 5.1b). For the formation of this kind of undulation, the environment must be large enough, and material must resist in

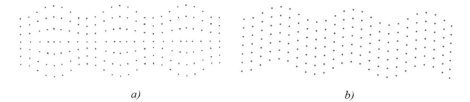

a)   *b)*

Fig. 5.1   Schematic illustration of undulation [7]. **a** Longitudinal undulation, **b** traverse undulation

the shear stress (i.e. solid environment only). Its velocity is affected by the shear modulus of elasticity [7, 17].

Each acoustic undulation can be characterized by different parameters. One of the most significant parameters is the acoustic pressure [17]:

$$p = \frac{F}{S} \tag{5.2}$$

where

p is immediate acoustic pressure (Pa),
F is an acoustic force (N), and
S is surface, on which acoustic force operate (m^2).

From the acoustic pressure, it is possible to express the acoustic performance [17]:

$$P = p \cdot S \cdot v \tag{5.3}$$

where

P is immediate acoustic performance (W),
p is acoustic pressure (Pa),
S is surface, on which acoustic pressure operate (m^2), and
v is the velocity of spreading of sound in the environment (m s^{-1}).

In Table 5.1 is shown the velocity of spreading of ultrasound in different environments. This velocity is affected by Poisson's number and density of material [9, 17].

Based on the acoustic performance, we can calculate intensity of acoustic field [17]:

$$I_\mathrm{a} = \frac{P}{S} \tag{5.4}$$

where

I_a is the intensity of acoustic field (power of ultrasound) (W m^{-2}),
P is acoustic performance (W), and

Table 5.1 Density and velocity of spreading ultrasonic waves in different environments [1, 17]

Matter	Density (kg m^{-3})	Velocity of spreading of longitudinal waves (m s^{-1})
Aluminium	2700	6320
Ferritic steel	7800	5920
Austenitic steel	8030	5200–5800
Nickel	8800	5630
Cast iron	7200	3500–5600
Siliceous glass	2600	5570
Porcelain	2400	5300–5500
Copper	8900	4700
Zinc	7100	4170
Brass	8100	3830
Gold	19,300	3240
Acrylic	1180	2680–2740
PVC	1400	2395
Lead	11,400	2160
Glycerine	1261	1920
Water at 20 °C	997	1483
Engine oil	870	1470
Chalcedony	2630	6800
Quartz	2650	7000
Jadeite	3430	8800
Olivine	3340	9000
Magnesia	3580	10,000
Corundum	4000	10,500
Obsidian	2376	5820
Garnet	4247	8510
Zircon	4596	3540

S is surface, on which it operates (m^2).

Based on the intensity of the acoustic field, we are a discerning ultrasonic field with low intensity (up to 5 kW m^{-2}), with medium intensity (5–100 kW m^{-2}) and with high intensity (over 100 kW m^{-2}) [16, 17].

5.2   Damping of the Ultrasound

The intensity of ultrasound is decreasing with increasing distance from the ultrasonic source (ultrasonic emitter). It is caused by absorption of the environment and by geometric factors in the environment, such as quarry, reflection, bending and dispersion of undulation. Therefore, concentrators and other components intended for transferring the ultrasonic energy are made of materials with low absorption of ultrasonic waves, such as titanium alloys, aluminium alloys and some kinds of properly treated steels [11, 12].

Geometric factors are affected by the heterogeneity of the environment, the presence of reflexive surfaces and foreign particles in the environment, as well as the wave length of ultrasonic undulation. Absorption is created due to the friction of oscillating particles, which cause the transformation of mechanical energy into thermal energy. Different materials have different values of absorption. Low absorption is typical especially for metals and alloys, such as aluminium, titanium and some steels. Medium absorption is typical for cast steel, copper, brass, bronze. A higher value of absorption has cast iron and hard polymers. The highest absorption is typical for materials such as rubber, soft polymers, porous materials, as well as lead and tin. Absorbed intensity can be calculated through the equation [12]:

$$I = I_0 . e^{-k.x} \tag{5.5}$$

where

I is intensity of wave (W m^{-2}),
I_0 is intensity of wave in position $x = 0$ (W m^{-2}),
e is Euler number (–),
k is Boltzmann constant (J K^{-1}), and
x is distance (m)

Generally, the solid matter has the lowest absorption, the liquid matter has a higher value of absorption and gasses have the highest absorption. For example, ultrasonic undulation with frequency 1 MHz, in distance 200 mm from the ultrasonic source, is reduced about 1% in steel (negligible), about 20% in water, and about 99% in the air (almost completely). We can observe that the reduction of ultrasonic intensity is linear for steel and water, but exponential for the air [18].

Absorbed energy is transformed into heat, which is manifested by warming. It can be observed especially in polymers; therefore, ultrasonic welding is suitable especially for the thermoplastics.

When the ultrasonic wave passes from one environment to another one, on the interface of these two different environments the quarry and/or reflection can occur. These phenomena are like the phenomena in optics. They also depend on the direction of the impact of ultrasonic undulation (perpendicular or oblique). If ultrasonic undulation hits the interface of environments in the perpendicular direction, part of undulation passes through this interface, and part of the undulation is reflected back to the ultrasonic emitter. When ultrasonic undulation hits the interface of environments under some angle (oblique impact), part of ultrasonic undulation passes through this interface, but under different angle (quarry), and part of the ultrasonic undulation is reflected back to initial environment and angle of reflection is equal to the angle of impact (similarly in optics). Amount of reflected undulation depends on materials of environments, as shown in Table 5.2. However, increasing of parallelism of impacting undulation to the environment interface (increasing angle between wave and normal vector of the interface), partial (or total—at high value of this angle) transformation of longitudinal waves into transverse (or surface) waves can occur [16, 17].

Each particle of the wave surface is a source of other undulation. Therefore, the wave can be bent if it interacts with the barrier (impermeable plate or hole). If the diameter of the barrier is much larger than the wavelength, shielding behind the barrier occurs. With decreasing of size of the barrier, bending of undulation begins. If the diameter of the barrier is comparable with the wavelength, perfect bend of undulation occurs, and the barrier has a very low disruptive effect [16, 17].

5.3 The Principle of Machining by Ultrasound

The principle of ultrasonic machining is in the abrasive effect of abrasives (usually scattered in the abrasive slurry, which circulates between the tool and workpiece) which ultrasound energy is delivered through a tool which vibrates the frequency of the ultrasonic waves.

Ultrasonic vibration occurs in a magnetostrictive or piezoelectric transducer (transducer converts electrical energy into mechanical–acoustic energy), which is powered by a high-frequency generator (high-frequency generator changes the input electrical power with frequency 50 Hz to the ultrasonic frequency, typically over 20 kHz). The oscillation thus created, although having a resultant ultrasound frequency, has too small an amplitude (1–100 nm) for machining purposes. Therefore, this vibration is transferred to the concentrator, where its amplitude is increased (to 25–90 μm). The concentrator is connected via an adapter to a working tool, which has a negative shape to the required profile of the workpiece. This schematic principle is shown in Fig. 5.2 [13].

Ultrasonic generators (high-frequency generators) are fed by electricity, while they are increasing the ordinary frequency of alternating current (AC) 50 Hz up to value 18–25 kHz. Properties of these generators have an influence on the functional and operating properties of the whole ultrasonic device. In present, semiconductor generators (thyristor or transistor) are the widest used. The ultrasonic generators themselves differ by their application—whether they are used to create ultrasound in a solid or liquid environment. These generators are connected with transducer [12, 17].

Ultrasonic transducers are converters, which convert electrical energy produced by ultrasonic generators into mechanical energy, which is more suitable for the purpose of mechanical machining. The widest deployed transducers are piezoelectric and magnetostrictive transducers [22]. Piezoelectric transducers utilize the piezoelectric effect of some dielectric crystalline materials (such as silicon), where electric discharge is arisen by mechanical deformation of piezoelectric material, or vice versa—if these piezoelectric materials are exposed to an electric field, mechanical deformation is induced in them. This effect is also called electrostriction, or a reversed

Table 5.2 Percentage losses of acoustic pressure caused by reflection between different environments [17]

Environment	Al	Mg	Cu	Brass	Ni	Steel	Pb	Hg	Hard rubber	Soft rubber	Glass	Acrylic	Porcelain	Engine oil	Water
Al	0	26	41	29	49	46	18	7	72	85	8	68	13	84	85
Mg	26	0	63	60	69	65	45	35	55	75	18	53	13	76	74
Cu	41	63	0	15	8	4	36	36	87	94	50	86	53	93	93
Brass	29	60	15	0	23	19	12	22	83	91	38	86	41	92	93
Ni	49	69	8	23	0	4	34	33	89	94	56	88	58	95	94
Steel	46	65	4	19	4	0	30	40	88	94	52	87	57	94	94
Pb	18	45	36	12	34	30	0	11	80	89	28	77	32	89	89
Hg	7	35	36	22	33	40	11	0	75	87	18	72	20	86	87
Hard rubber	72	55	87	83	89	88	80	75	0	50	67	7	40	30	31
Soft rubber	85	75	94	91	94	94	89	87	50	0	32	40	80	3	4
Glass	8	18	50	38	56	52	28	18	67	32	0	7	65	30	80
Acrylic	68	53	86	86	88	87	77	72	7	40	7	0	60	40	40
Porcelain	13	13	53	41	58	57	32	20	40	80	65	60	0	81	80
Engine oil	84	76	93	92	95	94	89	86	30	3	30	40	81	0	0
Water	85	74	93	93	94	94	89	87	31	4	80	40	80	0	0

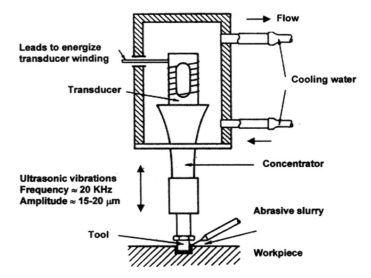

Fig. 5.2 Schematic principle of ultrasonic machining [13]

piezoelectric effect. Piezoelectric transducers have high energy efficiencies (approx. 90–96%), and they do not require any cooling. They are more easily adaptable for rotary operations, and they are more easily constructed than magnetostrictive transducers [12]. The magnetostrictive transducer utilizes the magnetostrictive effect of ferromagnetic materials (such as nickel). If these materials are exposed to a magnetic field, mechanical deformation (change of length) is inducted in them. The magnetic field is created by the transition of alternating current (AC) through coils retracted on the core. The magnetostrictive transducer is bulky because it must be air or water-cooled due to the generated heat. Generation of heat is caused by high energy losses which appear as heat. High energy losses cause decrement in the energy efficiency of a magnetostrictive transducer (approx. 55%) [2, 17].

Concentrator (sonotrode/horn) is a booster, whose purpose is to increase an input amplitude from the ultrasonic transducer, which has too low values for practical usage in the machining field, into several times higher value. This is obtained by the proper shape of the concentrator. Thus, a concentrator is a waveguide whose cross section is narrowing along the rotating axis. It usually has conic (Fig. 5.3a), exponential (Fig. 5.3b) or stepped shape (Fig. 5.3c). A significant factor for concentrators is the coefficient of amplification of amplitude. Concentrators thus serve to focus and concentration of ultrasonic oscillation in the low-frequency range of active ultrasound (up to 100 kHz) because of local increment in ultrasonic intensity. In high-frequency range (over 100 kHz), the mirrors (0.1–1 MHz) or lenses (over 1 MHz) are used for focusing. The concentrator should have good acoustic transmission properties, high fatigue resistance at high working amplitude, good corrosion resistance, good soldering and brazing characteristics, high strength to take screw attachments, etc.

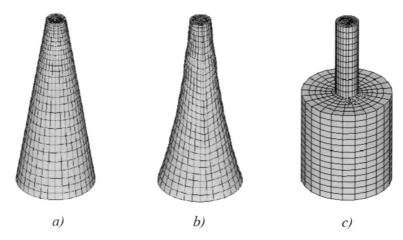

a)   *b)*   *c)*

Fig. 5.3 FEM models of concentrators [21]. **a** Conical, **b** exponential, **c** stepped

Therefore, it is usually made of Monel, titanium alloys, stainless steels, aluminium alloys, bronze and so on [2, 11].

The tool for ultrasonic applications is directly bonded with the concentrator. It should be joined by screws or by soldering or brazing. Many sellers of ultrasonic tools are selling concentrator as a part of the tool—they are permanently bound. The tool should be designed to provide the optimum amplitude of vibration at a given frequency. It should be of high wear resistance, high elastic and fatigue strength, good toughness and hardness, etc. Therefore, tools for ultrasonic machining are usually made of tungsten carbide (WC), silver steel, Monel and many others. Abrasive particles are made of very hard materials, such as polycrystalline diamond (PCD), cubic boron nitride (CBN), alumina (Al_2O_3), silicon carbide (SiC), boron carbide (B_4C), etc. Abrasive particles for USM are carried in coolant, for RUM they are bonded directly to the active part of the tool. Designs of the tools for RUM differ by their application. For example, for drilling the tool has a thin wall (Fig. 5.4a), because it is loaded only in the vertical axis and lower surface means lower cutting force and shorter machining time; and diamond particles are in the whole volume of the active part of the tool. For end milling, the tool has a thick wall (but it is still hollow due internal cooling) (Fig. 5.4b), because it is loaded in all directions; and diamond particles can be only on the surface of the active part of the tool. Ball milling cutters are not hollow, because otherwise, they would not be a ball in shape [2].

The principle of ultrasonic machining is based on the four different mechanisms of material removal. The first mechanism is the abrasive effect of abrasives that remove material by direct contact of abrasive particles with the workpiece surface. The second one is the microchipping which is caused by free abrasive particles. Another mechanism is the cavitation effect of suspension. The last one is the chemical effect

a)   *b)*

Fig. 5.4 CAD models of tools (with different concentrators) for RUM [8]. **a** Ultrasonic drill, **b** ultrasonic end mill

of the suspension. The predominant mechanism for material removing depends on the properties of a workpiece [2].

Traditional ultrasonic machining (USM) utilizes abrasive particles carried in the coolant (slurry) (see Fig. 5.5). As a coolant can be usually used water, but also oils, benzene, glycerol and their mixtures. Abrasive particles are usually made of polycrystalline diamond (PCD), cubic boron nitride (CBN), alumina (Al_2O_3) and other very hard materials. This slurry is fed between a tool and a workpiece. The tool is made of some soft material, such as brass or soft steel. The workpiece is harder and more brittle than the tool, and therefore, it is the workpiece that is being machined, not the tool. The tool oscillates in a vertical direction and is moving along this vertical

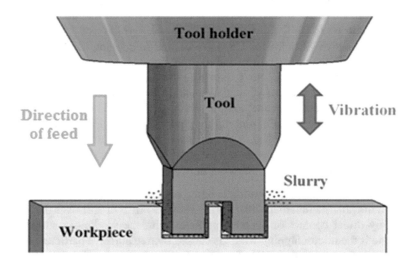

Fig. 5.5 Schematic illustration of USM [8]

axis. The shape of the tool is always negative to the hole. This type of machining is utilized for creating continuous as well as non-continuous holes with any kind of shape, not only circular. The tool is usually hollow for creating continuous holes because the lower surface means lower required energy and shorter machining time [10, 11].

Rotary ultrasonic machining (RUM) is a hybrid machining process, which combines traditional ultrasonic machining and diamond grinding. Tools for rotary ultrasonic machining oscillate not only in the vertical direction, but they also rotate around this vertical axis (see Fig. 5.6). Abrasive particles are not carried in the slurry, but they are bonded directly to the active part of the tool. These particles are usually made of synthetic diamond [15]. RUM can reach a higher material removal rate (MRR) than can be obtained by USM or diamond grinding. In addition, it can achieve a better quality of machined surface than the USM can do. Therefore, RUM is cost-effective machining technology for milling and drilling of hard and brittle materials, such as glass and ceramics. Tough and soft materials absorb ultrasonic energy and can seal pores on the active part of the tool. Difference between ultrasonic milling and drilling is based on the movement of the tool. During drilling, the tool is moving only in the vertical direction and therefore it is hollow with a thin wall. It is suitable for drilling circular continuous holes. However, during milling, the tool is moving

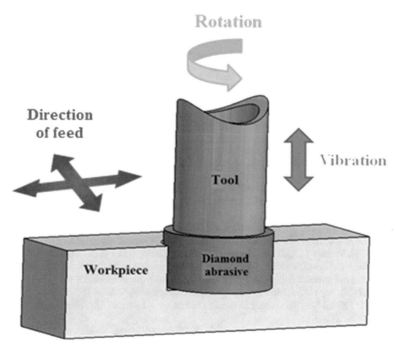

Fig. 5.6 Schematic illustration of RUM [8]

in all directions—vertical, as well as horizontal. Therefore, the tool for milling is stiffer, but it is also hollow—because of internal cooling [5, 6, 19].

In ultrasonic machining, no heat-affected zones are generated, and no chemical or electrical changes on the surface of the machined surface are present. Moreover, cutting forces are reduced. On the surface, low-pressure stress may be created, which increases resistance to cyclic fatigue. There is a challenge in traditional ultrasonic machining to achieve very small tolerances because the abrasive carried in the liquid coolant also wears the tool, thereby changing its size and geometry [3, 10, 20].

According to abovementioned information, the material affects the spreading of the ultrasonic waves. The damping of the undulation affects the tool material selection. The modulus of elasticity affects the ultrasonic energy absorption which affects the machinability of the workpiece material—soft materials absorb energy, while in brittle materials are spreading the cracks. To sufficient ultrasonic machining, the amplitude of the undulation has to be high enough. Devices for ultrasonic machining consist of ultrasonic generator (which increase frequency from 50 Hz to ultrasonic (20+ kHz) frequency), transducer (which change electric energy into mechanic one, while keeping the frequency), concentrator (which increase the amplitude from nanometres to micrometres) and tool (which actively provides the machining process). Correct selection of the tool material, workpiece material, ultrasonic set-up and machining parameters cause benefits during the cutting process.

References

1. Anderson OL, Liebermann RC (1966) Sound velocities in rocks and minerals. Geophysics Laboratory, Willow Run Laboratories, The Institute of Science and Technology, The University of Michigan. 7885-4-X, 195 p
2. Aspinwall DK, Thoe TB, Wise MLH (1998) Review on ultrasonic machining. Int J Mach Tools Manuf 38(4):239–255
3. Ceramic industry (2012) Machining ceramics with rotary ultrasonic machining. http://www.ceramicindustry.com/articles/machining-ceramics-with-rotary-ultrasonic-machining. Accessed 2 Oct 2012
4. Gregorová I, Ťažká MZ (2012) https://docs.google.com/viewer?a=v&pid=sites&srcid=ZGVmYXVsdGRvbWFpbnxyb3NpNpY292YxxneDo2YjMzOTNjZDY1MjEyYzkz. Accessed 2 Feb 2012
5. Hu P, Zhang JM, Pei ZJ, Treadwell C (2002) Modeling of material removal rate in rotary ultrasonic machining: designed experiments. J Mater Process Technol 129:339–344
6. Kocman K, Prokop J (2001) Technologie obrábění. Cerm, Brno. ISBN 80-214-1996-2
7. Kontroltech (2012) ultrazvuková metóda. http://www.ppkontroltech.sk/ndt/index.php?option=com_multicategories&view=categories&cid=7&Itemid=26. Accessed 3 Feb 2012
8. Kuruc M (2015a) Ultrasonic machining. Dissertation thesis, STU MTF, 172 p
9. Leitner B (2006) Nedeštruktívne skúšanie materiálov – kontrola ultrazvukom. In Techniky a technológie, Žilina
10. Maňková I (2000) Progresívne technológie. Publishers Vienela, Košice, pp 31–48. ISBN 80-7099-430-4
11. Mičietová A (2001) Nekonvenčné metódy obrábania. Publishers UNIZA, Žilina, 376 p. ISBN 80-7100-853-2

12. Mičietová A, Manková I, Velíšek K (2007) Fyzikálne technológie obrábania. Top trendy v obrábaní 5:22–48
13. Neelesh KJ, Vijay KJ (2001) Modeling of material removal in mechanical type advanced machining processes: a state-of-art review. Int J Mach Tools Manuf. 41:1573–1635
14. Rosičová M (2012) Fyzikálna podstata zvuku. https://sites.google.com/site/rosicova/studijne-materialy/multimedia/zvuk/fyzikalna-podstata-zvuku. Accessed 1 Feb 2012
15. Singh RP, Singhal S (2016) Rotary ultrasonic machining: a review. In: Materials and manufacturing processes, vol 31. Taylor & Francis Group, pp. 1795–1824. ISSN: 1042-6914
16. Švehla Š, Figura Z (1984) Ultrazvuk v technológií. Publishers Alfa, Bratislava. 528 p. MDT 534.321.9.000
17. Švehla Š, Abramov O, Chorbenko I (1986) Využitie ultrazvuku v strojárstve a metalurgií. Publishers Alfa, Bratislava, 320 p, 63-233-86
18. Tatar M (1962) Ultrazvuk ve strojírenství. Praha: Státní nakladatelství technické literatury, 121 p
19. Thoe TB, Aspinwall DK, Wise MLH (1998) Review on ultrasonic machining. Tools Manuf 38(4):239–255
20. USM (2012) Ultrasonic machining. http://www.ceramicindustry.com/articles/ultrasonic-machining. Accessed 3 Oct 2012
21. Wang F et al (2009) Development of novel ultrasonic transducers for microelectronics packaging. J Mater Process Technol:1291–1301. http://www.sciencedirect.com/science/article/pii/S0924013608002574
22. Zbojovská M (2013) Ultrazvuk. http://www.gjar-po.sk/~zbojovska8a/ULTRAZVUK.pdf. Accessed 1 Apr 2013

Chapter 6
The Application of the Ultrasound

Abstract The application of the ultrasonic energy into the production process positively affects every aspect of them. In machining, the ultrasound decreases the cutting
forces and temperatures and increases tool life a surface quality. It even allows to
machine hard and brittle materials. And it is not only machining, which finds out
benefits of ultrasonic undulation—even casting, forming, welding and soldering are
improved by ultrasound. The successful application of the ultrasound can be found
even in cleaning, localization, defectoscopy, metrology, medicine and many more.

Ultrasonic machining is used mainly for cutting hard and brittle materials (40 HRC
or more), which requires high precision of dimensions on the workpiece. The application is for the cutting of materials, excavating holes and grooves of different shapes
and sizes, deep hole drilling, threading production, and also for grinding, lapping and
polishing the glass and ceramics, in the manufacture of complex shapes of electrodes
for electro-erosive excavation, a complex-forming tool, in machining of ceramics,
composites, minerals, etc. [1, 12, 19].

   Besides ultrasonic machining, the ultrasound is also applied in welding, soldering,
brazing, cleaning, to improve formability, to refine the melt nucleation, in heat treatment, in defectoscopy, metrology, as well as in machining as secondary source of
energy to assist the primary source (ultrasonic-assisted turning, ultrasonic-assisted
milling, etc.) [1, 12, 17, 19].

6.1 Ultrasound in Metallurgy

Ultrasound can affect the process of casting and solidification in many ways. It has
an influence on the structure and properties of matter, which cause an improvement
of intensification of technological process. In the liquid phase, ultrasound causes
cavities and acoustic flows. These effects cause shuffle of liquid, improve its homog-
enization, increase diffusion processes, etc. Ultrasound can also decrease amount of
absorbed gases by liquid. These facts improve quality of ingots and casts, improve

© The Author(s), under exclusive license to Springer Nature Switzerland AG 2021   41
M. Kuruc, *Rotary Ultrasonic Machining*, Manufacturing and Surface Engineering,
https://doi.org/10.1007/978-3-030-67944-6_6

nucleation, recrystallization and growth of crystals, improve metallization of hard-wettable surfaces, improve spraying of liquid metal for manufacturing of metal dusts, improve the manufacture of pseudo-alloys and so on [17].

In ingot, which has been exhibited to ultrasound during solidification, fine regular grains, increase of homogeneity, decrease of liquation processes and due to cavitation also degassing can be observed. Influence of ultrasound depends mainly on the properties of the environment, volume and shape of liquid and performance of the ultrasonic source. Increasing of ultrasonic intensity makes degassing faster, but degassing affects the temperature of the liquid as well. In industry, degassing is performed by vacuum, but ultrasound is more effective. The most effective process of degassing is a combination of vacuum and ultrasound. The most significant parameter for ultrasonic-assisted casting is the efficiency of penetration of ultrasonic vibration. The most effective way to feed the ultrasound is to bring the vibration directly to flow of melted metal during continual casting. In addition, this method does not demand as high acoustic performance as some other methods. During vacuum-arc casting and electroslag casting, the most effective feed of ultrasound is to bring the vibration to the bottom of the stiffed ingot. Casting could be improved also by ultrasound and electromagnetism simultaneously [5, 17].

Structural changes caused by ultrasound in ingots lead to the change of mechanical properties. For example, due to ultrasound, strength and formability of steel have been increased, even in cryogenic conditions, as well as at higher temperatures. Strength and hardness of aluminium and cobalt have also increased, as well as hardness of tin and bismuth. Even in magnesium alloys, the ultrasound can soften the grains and improve formability in all temperature intervals. In copper alloys has been observed softening of structure, increasing of mechanical properties, as well as better corrosion resistance. Ultrasound also increases impact bending strength, increases purity and decreases liquation of alloying elements, even with considerably different density. Therefore, we should suppose that ultrasound can positively affect casting of majority of materials. In addition, ultrasound allows casting pseudo-alloys (composite alloys of mutually non-soluble elements, such as metal and oxide). In Fig. 6.1 is shown a

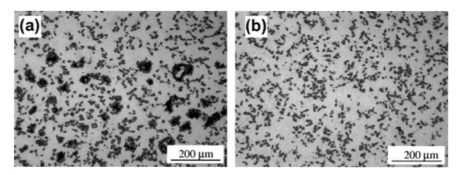

Fig. 6.1 Microstructure of Al-MMC casted [23]. **a** Without assistance of ultrasound, **b** with assistance of ultrasound

comparison of the microstructure of aluminium metal matrix composite (Al-MMC) reinforced by SiC particles and Zn-based filler metal manufactured by mechanical stir casting [23].

Fabrication of pseudo-alloys is usually performed by powder metallurgy. However, ultrasound can improve powder metallurgy as well. Presence of ultrasonic vibration during the creation of powder ensures finer and more uniform dust particles. It also increases the process productivity [17].

6.2 Ultrasound in Forming

The presence of ultrasound in solid metal causes changes in its mechanical properties and affects formability. Ultrasound decreases friction, energy loses, residual stresses, it increases velocity of forming, tool life, surface quality of components and enables to form materials, which are not suitable for forming in common conditions.

Material formed with assistance of ultrasound shows increasing of mechanical properties (strength, hardness) due to increasing of density (amount) of dislocations. However, during the operation of ultrasound, material is more formable and therefore for forming the lower level of stress (force) is required. It is caused by alternating stresses developed by ultrasonic waves and local increase of temperature (losses of ultrasonic energy). Alternating stress also causes that material will bend at a lower value of static stress, when the ultrasound is active [4, 17].

During drawing, ultrasonic vibration causes decreasing of drawing force and increase of trigger level of deformation per one draw. Therefore, lower number of repeating of the process and supporting operations was achieved when ultrasound was used, and this led to acceleration and cheapening of manufacture. During extruding, the decreasing of pressing force has been obtained, as well as decreasing of non-uniformity of deformations. Ultrasonic vibrations are usually fed to tool. During wire drawing of hard-deformable materials (such as Mo, Ta, W) to just a few micrometres thin wires, the ultrasound can be fed through lubrication liquid. This causes the object cleaned by ultrasound increases the surface quality of wire itself and decreases the wear of die [4, 17].

Force reduction by ultrasound is more effective with increasing of the ultrasonic amplitude. During extrusion of metal tubes with utilizing of the mandrel, reduction of mandrel's load is more effective, when ultrasound is fed directly to the mandrel (especially when mandrel rotates). There the ultrasound also causes increasing of the velocity of forming. Ultrasonic vibration decreases deformation stress even during tamping of steels. Tin, as well as copper, could be pressured at twice as higher level when compared to the process without ultrasound. Aluminium could be pressured even five times higher in comparison to drawing without ultrasound. However, the efficiency of ultrasound is decreasing with increasing of dimensions, level of deformation and usually also with increasing strength of deformed material. The highest influence of ultrasound on forming is visible in the forming of brittle materials, such as bismuth. Bismuth (Bi) cannot be formed under normal conditions. However, when

Fig. 6.2 Ultrasonically
assisted drawing [6]

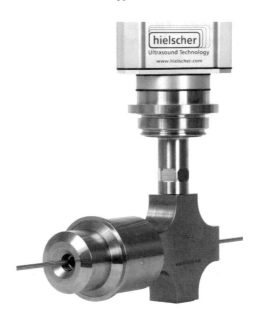

Bi is exhibited to ultrasonic vibrations, it can be formed [4, 17]. Therefore, ultrasonically assisted drawing of pipes, rods and wires began more commercial. A device
for a such process is shown in Fig. 6.2 [6].

Ultrasound positively affects the mechanical properties of formed material as
well. Strength of aluminium, copper and steel is increased. Microhardness of steel
is little increased, but for aluminium is decreased. Residual stresses are considerably decreased. Ultrasound also can decrease the surface roughness. Decreasing of
roughness is more obvious for soft materials, such as copper.

In the industry of forming, ultrasound is used especially for the drawing of metal
pipes, because ultrasound decreases pulling force, increases the degree of compression, increases geometrical accuracy and decreases the roughness of tubes. It causes
reduction of demanded operation, decreasing fabrication time (several times), and
increases the productivity of the process. Large application of ultrasound can be
found also in the drawing of thin wires from various metals [13, 17].

6.3   Ultrasound in Welding

Ultrasonic vibration can be used for welding of metals and polymers. Vibrations in
metals cause increasing of temperature, plastic deformation, internal stresses and
alternating shear forces. These phenomena accelerate mutual diffusion of welded
materials. The vibrations in polymers cause friction, increasing of temperature and

shuffle of welded materials. For welding of metals is recommended to use vibration in the direction parallel to the metal surface, but for welding of polymers is recommended to use vibration in the direction that is perpendicular to the polymer surface. Ultrasonic welding allows joining thin foils, sheets, wires, etc. Welding runs without additional material or cleaning. Properties of materials would not be changed, because there is a short welding time (fractions to a couple of seconds) and lower temperatures (lower power is required). It can weld electrically non-conductive material combinations (e.g. aluminium to glass). Welded materials do not demand input processing, and welding is easy to be automatized.

For welding of thin materials (few micrometres thick) are recommended to use higher frequency and lower amplitude (lower performance). Otherwise, the join could be ripped sooner than it is created. A very important parameter is also time. A too long time could cause outflow during welding of polymers, or cracks during welding of metals [17, 18].

Ultrasonic welding can be divided to spot welding, seam welding, circular welding and shaped welding. Pure metals are better weldable than alloys. Better weldability is typical also for ductile materials and materials with a lower melting temperature. Almost all combinations of metals can be welded, but due to the economic aspect, some of them are not utilized. For example, Al can be welded with Ag, Au, Be, Cu, Fe, Ge, Mg, Mo, Nb, Ni, Pd, Pt, Sn, Ta, Ti, W and Zr, but due to the economic aspect, the welding of Al with Pb, V and Zn is not recommended. There is a possibility to weld some soft materials to Si, glass, ceramics, etc. Hard-weldable materials could be welded, when thin interlayer (foil made of easy weldable material) is inserted (or steamed) to the welding interface. For welding of the polymers, the weld is easier to make in hard polymers with a similar melting point. Soft polymers have higher absorption of ultrasonic energy [17, 18].

Ultrasonic welding of metals is utilized especially in the electrotechnical and automotive industry. Ultrasonic welding of polymers can be utilized even to join textile made of synthetic thermoplastic fibres. Ultrasonic welding device has high energetic effectiveness. In comparison with resistance welding machine, ultrasonic welding device has 25 times higher efficiency. During welding of polymers, efficiency could be increased by creating projections as concentrators of ultrasonic energy on welding parts. Ultrasonic welding device could be used, e.g., to welding, riveting, forcing-in and edging metal and polymer parts with different materials. Due to high energy efficiency, ultrasonic welding machines have relatively low dimensions (can be constructed even in compact form). They also can have several welding heads [18].

Except welding, ultrasonic energy could be used in soldering and brazing. It is utilized especially for joining of aluminium because when this metal is exposed to air, it creates passivation layer of hard-meltable Al_2O_3 on its surface. Ultrasonic vibrations cause cavities in melted solder, which violate this oxidic deposit. The surface of the base metal becomes a little rougher and therefore there is observed improved adhesion of a joint. Ultrasonic soldering is also used for soldering of glasses and ceramics. Solder is melted by additional devices, such as burner and furnace. Ultrasonic vibrations are used only to improve the process. The process is faster, the

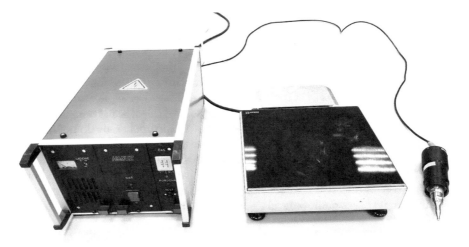

Fig. 6.3   Device for ultrasonic soldering

joint has better properties, there is decreased request for fluxes and it is possible to solder materials in an easier way, even those that are really hard-solderable without ultrasound. Therefore, ultrasonic vibrations increase economic effectiveness as well as productivity [17, 20]. Owen with ultrasonic generator and sonotrode, using for soldering, are shown in Fig. 6.3.

6.4   Ultrasound in Machining

Ultrasound in machining can be used to assist in conventional machining processes (such as ultrasonic-assisted turning, ultrasonic-assisted milling) [7, 8, 10], or directly in the machining of materials (ultrasonic machining, rotary ultrasonic machining) [16]. Thanks to ultrasound the hard-machinable materials could be machined more effectively. Ultrasound even allows to machine materials, which cannot be machined by other methods. Ultrasonic machining is utilized especially for hard and brittle materials. These materials usually have great chemical and thermal resistance, structural and optical homogeneity, and other valuable properties, which make them irreplaceable in some applications. However, these unique properties cause their hard-machinability. Conventional method either cannot machine them, or achieved results are unacceptable (creation of cracks, etc.), or the whole process is not economically effective (too long machining time, too short tool life).

   Ultrasonic machining (USM) is utilized in drilling, dividing (cutting), creating complex shapes of machined surface, etc. especially for hard and brittle materials (such as glass, ceramics, minerals and silicon), where we can reach surface roughness parameters comparable with surface roughness parameters achieved by grinding.

Both continuous and blind holes can be created. This method is effective especially for machining of hard and brittle materials. Productivity, as well as the surface quality, depends on amplitude, frequency, size of abrasives, abrasive concentration, slurry circulation, materials of the workpiece, abrasive and tool, type of coolant, downforce of the tool, and shape and dimensions of the tool. With increasing of amplitude, tool's downforce and size of abrasive are increasing productivity and reached the rougher surface. Tool wear affects the accuracy of blind holes much more than affects the continuous ones. Tough materials absorb more acoustic energy, and therefore, the process is slowed or totally stopped. The machined surface is without cracks, and under surface are present compressive residual stresses, which have a positive character. This method is utilized especially in optics for production of lenses and in electronics to manufacture semiconductors [12, 17].

Productivity can be increased by rotary ultrasonic machining. There is diamond abrasive spread directly on the active part of a tool, which rotates around its own axis, as shown in Fig. 6.4. This method can be applied for higher depths. In addition, tool wear is lower and surface quality better. Rotary ultrasonic machining represents a unique combination of ultrasonic machining and diamond grinding (DG) into the hybrid method, which surpasses both methods (USM & DG). In comparison with traditional ultrasonic machining, higher material removal rate (MRR) is achieved there, hand in hand with a smoother surface and better geometric precision [16].

Ultrasound in machining can be used also to support other machining methods. For example, to machine hard alloys, electrochemical machining supported by ultrasonic vibrations could be used. It causes increasing of productivity and decreasing of tool wear (the tool is also cathode). However, it could be applied only for machining of electrically conductive materials. Ultrasound can also improve mechanical methods of machining. It decreases cutting forces during turning, milling, drilling, grinding, shaping, countersinking, threading, etc. It causes increasing of productivity, accuracy, as well as it prolongs the tool life, and on the other side, it decreases roughness.

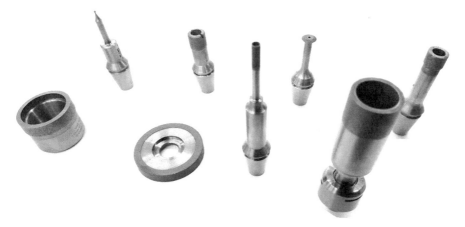

Fig. 6.4 Different shapes of tools for rotary ultrasonic machining

It is utilized especially for machining of strong, tough, creep-resistant and hard-machinable alloys (such as Ni alloys). During ultrasonic-assisted turning of brittle materials, the cutting force and surface roughness parameters were decreased, build-up-edges (BUE) were not created and tool life has been increased up to seven times. During ultrasonic-assisted threading of tough and hard materials, the friction, torque and cutting force have been decreased. It is applied especially for the manufacturing of external thread (especially for threads with small dimensions). Tool life could be increased up to ten times. During ultrasonic-assisted grinding, polishing and lapping, the cutting force has been decreased and improved quality of surface was achieved. In addition, productivity has been increased up to five times. During ultrasonic-assisted drilling and ultrasonic-assisted reaming of stainless steels and Cu and Al alloys, productivity, accuracy and tool life have been increased, and torque, axial force and roughness have been decreased. Ultrasonic-assisted drilling is especially suitable for drilling into thin sheets. Comparison of ultrasonic-assisted drilling and conventional drilling is shown in Fig. 6.5. Inconel 738-LC has been used as a workpiece. Following parameters have been used: feed rate 0.8 mm/s and spindle speed 250 rpm. During ultrasonic-assisted drilling has been used frequency 21 kHz and amplitude 10 μm. The diameter of the hole is 5 mm [3, 17].

The difference in set up between conventional milling, ultrasonic-assisted milling and ultrasonic milling (rotary ultrasonic machining) is in the used cutting tool. During rotary ultrasonic machining, there is the tool in form of grinding tool (often with concentrator) mounted into ultrasonic tool-holder, as shown in Fig. 6.6. During ultrasonic-assisted milling, there is also used ultrasonic tool-holder, but conventional milling cutter is mounted into it. There are preferred smaller diameters of the cutters, which going to the higher diameter of the shrank via cone (cone can be used as concentrator). During conventional milling is used conventional milling cuter mounted into conventional tool-holder [9].

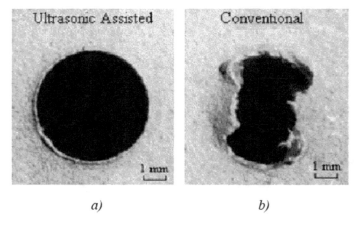

a)   *b)*

Fig. 6.5 Burr formation during drilling of nickel alloy [3]. **a** Ultrasonic assisted drilling, **b** conventional drilling

Fig. 6.6 Ultrasonic tool mounted into ultrasonic tool-holder [9]

6.5 Other Applications of the Ultrasound

The ultrasound can be applied also for cleaning, deburring, surface hardening, heat treatment, chemical-thermal treatment, coating, in defectoscopy, localization (sonar), metrology, medicine, etc.

In industry, the ultrasound is commonly applied for removing greasy and mechanical impurities from the surfaces. Devices for ultrasonic cleaning are relatively simple and available, and cleaning has great effectiveness and quality [on the surface stay only 0.5% of impurities after the process (other cleaning methods remove only 20 to 80% of impurities)]. This process is cheap, fast and easy to make it automated. Ultrasonic cleaning is used especially for complex and small components. The process is based on cavitation and erosion effect. Cavitation effect has the biggest influence not only for cleaning but even in other applications of ultrasound. Also, the cleaning liquid can influence the cleaning process (it should chemically react with impurities, but not with the base material). Increasing of intensity of the process could be achieved by increasing hydrostatic pressure, liquid temperature, ultrasonic frequency, etc. Ultrasonic cleaning is utilized in dental and medical sectors, in ophthalmology and optics, in jewellery and watch-making branch, in laboratory, industry, workshops and services, etc. Ultrasonic cleaning can be used on metals, as well as non-metals, and even on their combination [17, 21] (Alibaba [2]).

Ultrasonic deburring is applied especially for small components. This process could be even more effective when small abrasive particles are present in the liquid. Quality of surface and productivity is increased when abrasive is present. This process can be used for metals as well as for non-metals.

More than anything else, the ultrasonic diamond smoothing is utilized for surface hardening. Ultrasonic vibrations enhance the process of plastic deformation, and lower static forces are required. The surface layer has lower roughness and higher tensile strength, impact bending strength, fatigue and abrasion resistance. Creation of microcracks on the surface is avoided. Utilized devices are relatively simple and reach high productivity. This method can be used only for plastic hardenable materials, such as steels and aluminium alloys. Life of the dies processed by ultrasonic hardening could be increased up to 20 times [15, 17].

Ultrasonic vibrations in process of heat treatment of metals accelerate the process itself. Diffusion is faster and homogenous, it penetrates deeper, especially at higher temperatures, and the phase recrystallization is faster, too. Speed of ageing and tempering can be increased up to 100 times. Particles of arming phase are segregated primarily inside of grains of a solid solution and not on the borders. Mechanical properties of treated components are improved. Higher hardness could be achieved, as well as impact bending strength. Even quenching is positively improved by ultrasound. The drain of heat is more rapid, and therefore, the cooling effect is more intensive. Hardness, re-hardenability and surface quality are increased.

Diffusion is accelerated in chemical–thermal treatment as well. It increases the productivity of the process. It is utilized especially for carbonization, nitrification and boridification. Depth of diffusion is increased too. Hardness, ductility and abrasion resistance of the surface layers are twice as high as in the process without ultrasound.

Even electrochemical processes, such as coating, are accelerated. The coated layer is homogenous, and its adhesion is increased. Coated metal has a finer structure and is less porous. Process of the coating by copper with the assistance of ultrasound is up to 15 times faster. Hardness and strength of the nickel layer are increased. The hardness of the chromium layer is also increased. Coating by cadmium is up to 10 times faster. The surface of the brass layer is glossier. Current density can be increased up to 15 times during coating by zinc. Ultrasound is also used for coating by silver, gold, tin, etc. [17, 22].

Ultrasonic waves could be also used for three-dimensional acoustic manipulation (acoustic levitation) of small objects made of different materials, such as metals, polymers, wood and water drops. The background was described by Whymark in 1975 in publication *Standing waves levitation*. The objects could move only in one direction. 2D standing waves manipulation was described in 2013 by two individual teams of authors: Foresti et al. and Kono et al. [11]. In the same year (2013) the 3D standing waves manipulation was also described by Ochiai et al. in the publication *Three-dimensional mid-air acoustic manipulation by ultrasonic phased arrays* [14]. Their device for acoustic manipulation is shown in Fig. 6.7. It consists of four speakers, which create phased array.

To sum up, the application of ultrasonic energy into the production process positively affects every aspect of them. Via rotary ultrasonic machining, it allows effective machine hard-machinable materials. And if ultrasound assists in conventional machining, benefits appear here as well. And it is not only machining, which finds out benefits of ultrasonic undulation—even casting, forming, welding and soldering are improved by ultrasound. Beside production processes, the ultrasound is beneficial even in cleaning, localization, defectoscopy, metrology, medicine and many more. It seems everything could be improved by ultrasound.

Speaker   Speaker

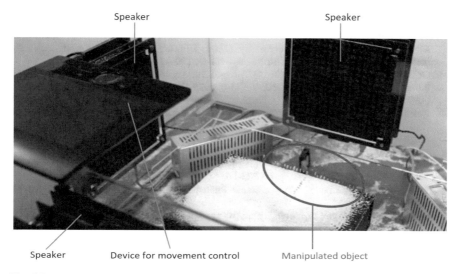

Speaker   Device for movement control   Manipulated object

Fig. 6.7   Acoustic manipulation [14]

References

1. Alexplus (2013). Development and application of the ultrasonic technologies in nuclear engineering. http://www.alexplus.ru/En/Papers/Development_and_Application_of_ the_Ultrasonic_Technologies_in_Nuclear_Engineering.html. Accessed 20 Aug 2013
2. Alibaba (2013) Professional industrial ultrasonic cleaner. http://www.alibaba.com/product-gs/ 674372014/Professional_Industrial_Ultrasonic_Cleaner_DR_LP80.html. Accessed 16 Sept 2013
3. Azarhoushang B, Akbari J (2007) Ultrasonic-assisted drilling of Inconel 738-LC. Int J Mach Tools Manuf 47(7–8):1027–1033
4. Djavanroodi F, Ahmadian H, Koohkan K, Naseri R (2013) Ultrasonic assisted-ECAP. Ultrasonics 53(6):1089–1096
5. Haghayeghi R, Kapranos P (2013) Direct-chill casting of wrought Al alloy under electromagnetic and ultrasonic combined fields. Mater Lett 105:213–215
6. Hielscher (2020) Hielscher ultrasound technology. Ultrasonically assisted drawing of wires, pipes and profiles (UAD). https://www.hielscher.com/ultrasonically-assisted-drawing-of-wires-pipes-and-profiles-uad.htm. Accessed 22 Sept 2020
7. Kuruc M, Zvončan M, Peterka J (2013a) Comparison of conventional milling and milling assisted by ultrasound of aluminum alloy AW 5083. In: IN-TECH 2013: proceedings of international conference on innovative technologies, Budapest, Hungary. Faculty of Engineering University of Rijeka, Rijeka, pp 177–180. ISBN 978-953-6326-88-4
8. Kuruc M, Zvončan M, Peterka J (2013b) Investigation of ultrasonic assisted milling of aluminum alloy AlMg4.5Mn. In: Annals of DAAAM and proceedings of DAAAM symposium: 24th DAAAM, Zadar. ISSN 2304-1382
9. Kuruc M (2015) Ultrasonic machining. Dissertation thesis, STU MTF, 172 p
10. Kuruc M, Šimna V, Vopát T, Peterka J (2016) Investigation of ultrasonic assisted milling of nickel alloy monel. In: Team 2016: proceedings of the 8th international scientific and expert conference, Trnava, Slovakia, AlumniPress, pp 47–52. ISBN 978-80-8096-237-1
11. Manipulation (2014) 3D standing waves manipulation. http://m.mojevideo.sk/?s=view&id= 2l55. Accessed 1 Jan 2014

12. Maňková I (2000) Progresívne technológie. Publishers Vienela, Košice, pp 31–48. ISBN 80-7099-430-4
13. Mordyuk BN, Mordyuk VS, Buryak VV (2004) Ultrasonic drawing of tungsten wire for incandescent lamps production. Ultrasonics 42(1–9):109–111
14. Ochiai Y, Hoshi T, Rekimoto J (2014) Three-dimensional mid-air acoustic manipulation by ultrasonic phased arrays. PLoS ONE 9(5):5 p, e97590. https://doi.org/10.1371/journal.pone.0097590
15. Rusinko A (2011) Analytical description of ultrasonic hardening and softening. Ultrasonics 51(6):709–714
16. Singh RP, Singhal S (2016) Rotary ultrasonic machining: a review. In: Materials and manufacturing processes, vol 31. Taylor & Francis Group, pp 1795–1824. ISSN: 1042-6914
17. Švehla Š, Abramov O, Chorbenko I (1986) Využitie ultrazvuku v strojárstve a metalurgií. Publishers Alfa, Bratislava, 320 p, 63-233-86
18. Turňa M (1989) Špeciálne metódy zvárania. Publishers Alfa, Bratislava, 384 p. ISBN 80-05-00097-9
19. USM (2012) Ultrasonic machining. http://www.ceramicindustry.com/articles/ultrasonic-machining. Accessed 3 Oct 2012
20. Wei J, Deng B, Gao X, Yan J, Chen X (2013) Interface structure characterization of Fe36Ni alloy with ultrasonic soldering. J Alloy Compd 576:386–392
21. Woodard (2013) Ultrasonic cleaning—a tool in Woodard's cleaning arsenal. http://www.woodard247.com/restoration/case-studies/15/ultrasonic-cleaning-a-tool-in-woodards-cleaning-arsenal#.Ujcqv6y2auA. Accessed 17 Sept 2013
22. Wu MH et al (2010) Preparation and characterization of nano Ni-TiN coatings deposited by ultrasonic electrodeposition. J Alloys Comp 490(1–2):431–435
23. Xu Z, Yan J, Shi L, Ma X, Yang S (2011) Ultrasonic assisted fabrication of particle reinforced bonds joining aluminum metal matrix composites. Mater Design 32(1):343–347

Chapter 7
Machinable Materials

Abstract The ultrasound is beneficial in the machining of a wide range of materials. However, the device for rotary ultrasonic machining is not cheap. It can reach the full potential when materials are machining, which are not easy machinable by other methods. Those materials include very hard materials, such as technical glass, advanced ceramics and minerals. The application is focused on the processing of synthetic diamond, cubic boron nitride, alumina, zirconia, carborundum, etc. Beside mentioned hard and brittle materials, ultrasound in machining can be used for machining of chromium-cobalt, titanium, wolfram carbide, graphite, martensite, ferrite, vanadium, stainless steel, composites reinforced by aramid/carbon/glass fibres, and many more. Such a range of machinable materials makes rotary ultrasonic machining attractive in sectors, where machining of different kinds of unique materials is needed.

Ultrasonic machining is proper especially for precise machining of hard and brittle materials. Hard materials are typical by high electron density and high bond covalence. Generally, with increasing of hardness, the brittleness is increasing as well [153]. These kinds of materials are used for specific applications due to their unique properties [151]. Usually, there is a demand for high accuracy. In many applications, they are irreplaceable [76]. For example, one of such applications is the nanoindenter for measurement of hardness, modulus of elasticity, yield strength, fracture toughness, scratch hardness and wear properties [24]. There is a diamond tip needed with exactly specified dimensions. If this indenter was not made of diamond, its applications would be limited. And if it had non-normalized dimensions, its results would have been of no use at all [47]. Therefore, there was investigated such materials that are characterized by their high hardness [71].

Diamond and CBN have significantly higher hardness than the other materials. For clarification, hardness equalling to 1 GPa corresponds to approx. 100 HV (there is a direct correlation) [10]. A typical application of rotary ultrasonic machining is in the dental industry in the manufacturing of dental implants made of hard materials, such as zirconia (hardness 12.3 GPa), lithium disilicate (hardness 5.8 GPa), promysan, etc. These materials have to be hard enough; otherwise, these implants cannot be used for

© The Author(s), under exclusive license to Springer Nature Switzerland AG 2021
M. Kuruc, *Rotary Ultrasonic Machining*, Manufacturing and Surface Engineering,
https://doi.org/10.1007/978-3-030-67944-6_7

biting. And they must be accurate enough; otherwise, they could be uncomfortable for the proprietor [149].

Another example of the application of hard materials with defined shape is the fabrication of welding tools for friction stir welding (FSW) process of advanced materials. Materials such as magnesium and aluminium alloys can be easily welded by the tool made of high-speed steel (HSS) [50]. However, such a tool cannot be used for welding stainless steels (SS), and nickel, cobalt and titanium alloys, because HSS does not have sufficient hardness and it has low thermal stability. Welding temperature at FSW is slightly lower than the melting temperature of welded material. HSS tool would be tempered at higher temperatures and losing its mechanical properties. Therefore, welding tools are made of polycrystalline cubic boron nitride (PCBN) for applications when materials with higher mechanical properties and higher melting point have to be welded [20]. These tools usually have a complex shape, which cannot be manufactured by powder metallurgy (due to volume retraction during sintering is not possible to achieve some shapes) [51]. Such a tool allows using higher welding speed and higher tool revolution when it is used for welding of softer materials [49, 53].

Following chapters are focused on selected hard and brittle materials, their properties and machinability via rotary ultrasonic machining. Closer look will be focused on synthetic diamond, cubic boron nitride and alumina. Also, there will be described zirconia, silicon carbide and other materials, such as aluminium nitride, silicon nitride, borosilicate, silicon oxide, silicon, boron carbide, etc. There is also recommendation of machining parameters for those materials.

7.1 Synthetic Diamond

Synthetic diamonds (SD) are produced at high temperature (1500 °C) and high pressure (5 GPa) [30]. They consist of pure carbon which is crystalized to isotropic 3D form [38]. They can be manufactured as monocrystalline diamonds (MCD), or polycrystalline diamonds (PCD). Properties of SD depend on their purity and quality [154]. Therefore, manufacturers usually divide SD into grades according to their purity [29]. The highest grades of SD have similar properties (such as hardness and thermal conductivity) that are even superior to the majority of naturally formed diamonds [145]. SD are usually utilized as abrasives for varied methods of machining [77].

Diamonds are the hardest known material [71]. They have the value of 10 in the Mohs scale of mineral hardness [85]. Their hardness is about 10,000 HV (100 GPa), depending on the purity of SD, crystalline perfection and orientation (best orientation is in [111] direction) [73]. Diamonds are chemically inert. Pure diamond is electrical insulator; however, impurities can affect its properties, such as hardness, strength, toughness and conductivity [156]. Elastic modulus of a diamond is 1140 GPa. Its density has the value approx. 3515 kg m^{-3} [11]. Diamond is very good thermal conductor. It has the highest thermal conductivity of all solid materials [2000–3000 W/(m K)] [63]. This value is higher than the value of the thermal conductivity of any metal [23]. They have also a low coefficient of thermal expansion $(8 \times 10^{-7} \text{ K}^{-1})$ [9, 13, 144].

Due to affinity of carbon to iron at high temperatures, diamond dissolves in iron and forms iron carbides, and therefore, it is inefficient in cutting ferrous materials, such as steel [35, 84]. Another disadvantage of diamonds is their oxidation at elevated temperatures (above 800 °C), and therefore, they cannot be effectively used at high cutting speeds [44, 72]. Synthetic diamonds are usually utilized to machine non-ferrous materials [15]. They can be applied also in high-power CO_2 lasers as gyrotrons or output window; or in high-power radiation sources as a diffraction grating (synchrotrons) [32, 75, 142]. In electronics, SD can be used as semiconductor (diode), transistor, or radiation detection device [4, 18, 33, 39, 79].

Diamond has a metastable structure (cubic lattice). When it gets energy (heat), its structure stabilizes and the diamond transforms into graphite (hexagonal lattice). Diamond, as well as graphite, consists of pure carbon. However, their properties are markedly different. Graphite is soft (value approx. 1.5 in Mohs scale of hardness). It is caused by the weak atomic bonds in the vertical direction of the lattice (between planes). In horizontal direction affects strong covalent bonds; however, between the layers themselves, it affects only weak van der Waals forces. This also causes anisotropic properties. Graphite has also lower density than diamond—only 2160 kg m^{-3}. Graphite is often utilized as a lubricant. Unlike diamonds, graphite is not transparent—it has black colour [6, 17, 64].

One of the producers of synthetic diamonds at relatively low prices is Chinese company Changsha 3 Better Ultra-hard Materials, Co. Ltd. The biggest monocrystalline diamonds, which they are offering is a block-shaped MCD with dimensions 6.5 $\times$ 6.5 $\times$ 1 mm. Those diamonds could be applied for processing the hard metal-bond and electroplated tools, grinding apparatus, geological drilling, etc. [1].

Company Ultra-hard China offers large single crystal diamonds with size 4 mm in the form of cubo-octahedral shape. Controlled growth is used during high-pressure, high-temperature (HPHT) synthesis. Their application is in dressing tools, jewellery processing tools, wire drawing dies, mechanical processing, and precision tools for precious stones and noble metals, etc. [148].

The synthetic diamonds for technical applications are transparent, but yellow in colour. Synthetic diamond used in jewellery is transparent, colourless, bigger, polyhedral and much expensive.

Because diamond is considered as the hardest known material, problems with its machining could be expected. It can be machinable when the processing of brilliants, which are machined by the diamond grinding. Rotary ultrasonic machining also utilizes diamond tool similar to small grinding wheels; however, cutting speed is markedly lower, the feed rate is higher, and depth of cut (DOC) is also higher in comparison with diamond grinding. Theory of machining highlights that material of the tool must be harder than the material of the workpiece in mechanical machining [8, 40–43]. And this condition cannot be fulfilled when the hardest material is used as the workpiece. Anyway, diamond is machinable by RUM. The machining process is stable and low machine loads appear in the process. However, the tool wear is obvious even by the naked eye. Diamond (in most of its shapes) has sharp edges in interfaces of its surfaces. And because diamond is very effective cutting material, a workpiece cuts the tool as well as the tool cuts a workpiece. If a workpiece had rounded or chamfered edges, the tool should be less damaged.

There were performed research in rotary ultrasonic machining of the monocrystalline diamond [55, 56] using an ultrasonic milling cutter with diameter 24 mm. There were obtained following dependences during machining of MCD: with increasing of each machining parameter (cutting speed, feed rate, DOC), torque is slightly increasing. Each parameter change had almost the same effect. Machine load in Z-axis was almost not affected by the cutting speed and DOC. However, there was recorded a big influence of the feed rate. With its increasing, machine load is also increasing, as shown in Fig. 7.1 [58, 59].

Tool wear causes braking of the edge of a tool. It makes the ratio of material removed from a workpiece to material removed from the tool (i.e. grinding ratio) approximately only 1:5.6, as shown in Fig. 7.2. It means the material was removed from tool 5.6 times faster than from a workpiece (depending on the machining parameters). It could be caused by a solid monocrystalline structure of the diamond workpiece, while diamond particles in a tool are bonded by brass matrix. According to achieved results, the following influences could be inferred: increasing the feed rate and DOC cause increasing of the tool wear. However, increasing the cutting speed causes a decrease in the tool wear. It means, in terms of tool wear, the process is better in conditions more similar to diamond grinding. However, grinding is more limited by the complexity of the component shape. Therefore, it still cannot replace RUM in some applications. Low surface roughness parameter Ra was obtained during machining monocrystalline diamond. Depending on the machining parameters, its value was in range 0.02–0.54 μm [55].

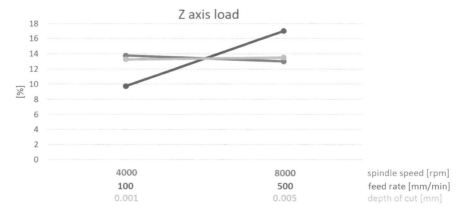

Fig. 7.1 Influence of machining parameters on machine load at machining of MCD [55]

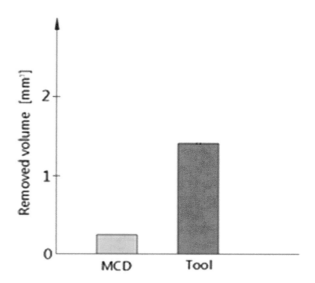

Fig. 7.2 Volume removed from MCD and volume removed from the tool [55]

According to performed experiment, there is recommended following machining parameters during RUM of MCD for the cutting tool Ø 24 mm: high cutting speed (many ultrasonic tool-holders are limited to 8000 rpm in spindle speed, but there are available the ones, which could be used up to 40,000 rpm) in terms of low tool wear; high depth of cut (5 μm) in terms of higher material removal rate; and lower feed rate (100 mm/min) in terms of machine load in Z-axis; however, when machine tool is rigid enough, there can be used higher feeds.

7.2   Cubic Boron Nitride

Cubic boron nitride (CBN) is the second hardest material. In nature, only hexagonal boron nitride is present, because the hexagonal structure is more stable than the cubic one [141]. The hexagonal form of boron nitride has similar properties as graphite, and it is also utilized as a lubricant. CBN is fabricated similarly as synthetic diamonds— at high temperature and high pressure [152]. Hexagonal boron nitride powder is exposed to pressure in the range from 4 to 18 GPa and sintered at a temperature from 1500 to 3230 °C [150]. In comparison to diamond, CBN has higher thermal and chemical stability. It has no affinity with iron; therefore, it can be used for machining of hardened steels and tool steels [25, 45, 48].

The hardness of CBN in the Mohs scale reaches the value between 9.5 and 10. It corresponds with hardness about 70 GPa (7000 HV). Resultant hardness is affected also by the grain size (in polycrystalline structure). Polycrystalline cubic boron nitride (PCBN) with grain size in the order of 10 nm could reach even higher hardness than diamond [78, 155]. However, bigger grain size can cause decreasing of resultant hardness even to 40 GPa (4000 HV). CBN is thermally resistant to decomposition at temperatures: 1300 °C in air and 1600 °C in a vacuum. The density of CBN is similar to diamond's one—3480 kg m^{-3}. Elastic modulus reaches the value of 850 GPa. Thermal conduction of CBN is also high—1300 W/(m K), and its coefficient of thermal expansion is 3.5×10^{-6} K^{-1} [65, 71].

CBN is widely applied as abrasive, especially for machining of hardened ferrous alloys [147]. It is also utilized as a part of high-temperature devices, as a heat spreader, for X-ray membranes, for light-emitting diodes (LED) or lasers [28, 46, 146].

Polycrystalline cubic boron nitride (PCBN) is manufactured by Chinese company Changsha 3 Better Ultra-hard Materials, Co. Ltd. They provide different shapes and sizes of specimens. In Fig. 7.3, there is the energy-dispersive X-ray(EDX) spectroscopy microanalysis of PCBN sample, where the distribution of chemical elements is shown [1].

According to the EDX analysis, it is obvious that PCBN consists of CBN grains in alumina matrix doped by cobalt. For better understanding, the obtained chemical composition is recorded in Table 7.1.

Laser beam machining is an effective method for processing PCBN [54]. However, it can cause unwanted affection in the surface layer. There were performed research in rotary ultrasonic machining of polycrystalline cubic boron nitride [55] with an ultrasonic milling cutter with diameter 24 mm. During machining of PCBN, a load increment was observed. Moreover, the character of material caused its sticking on the tool. It caused additional load increasing. Therefore, loads during machining were measured in the middle of a process and near the end of the process. Between every single machining of an area, the tool was cleaned [52, 53, 57].

According to the measured machine outputs, there can be obtained following dependencies: with increasing cutting speed, torque is increasing, as well as machine load in Z-axis. Feed rate has similar behaviour as cutting speed—torque is increasing, as well as machine load in Z-axis. With increasing DOC, torque is slightly increased, and the load in Z-axis is markedly increased. Machine load in Z-axis was the most restrictive parameter. However, torque was also very high—a tool with a lower diameter could be damaged at such conditions. The longer time the tool was machining, the more material of a workpiece was stuck on the tool and machine loads were more increased. Recorded influences of examined machining parameters on the torque and machine load in Z-axis are shown in Figs. 7.4 and 7.5, respectively [59, 60].

The tool wear was relatively significant. There were achieved the ratio of material removed from a workpiece to material removed from a tool (i.e. grinding ratio) approximately only 2.654:1, as shown in Fig. 7.6. According to the achieved results, the following influences could be inferred: increasing feed rate causes increment in tool wear. DOC has inverted influence—increasing DOC causes lower tool wear. This fact is caused by a lower amount of workpiece sticking on the tool at constant machined volume. Influence of the cutting speed is different. In the beginning, increasing of cutting speed caused slightly decreasing of tool wear, but additional

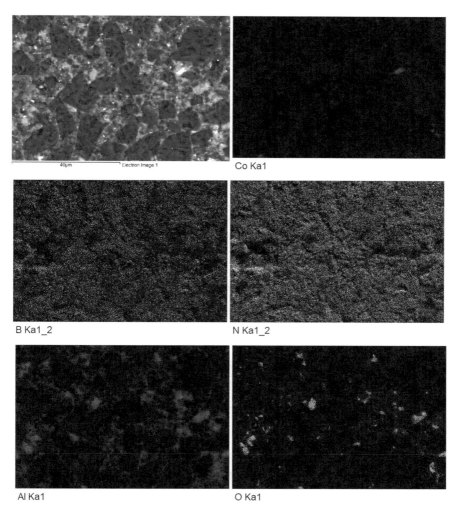

Fig. 7.3 Chemical distribution of elements in PCBN specimen [55]: evaluated surface (gray); cobalt distribution (violet); boron distribution (green); nitrogen distribution (yellow); aluminium distribution (red); and oxygen distribution (turquoise) respectively

Table 7.1 Chemical composition of PCBN [55]

PCBN	B	N	Al	O	Co
wt%	46.99	40.58	6.26	4.68	1.49
at. %	55.77	37.18	2.98	3.75	0.32

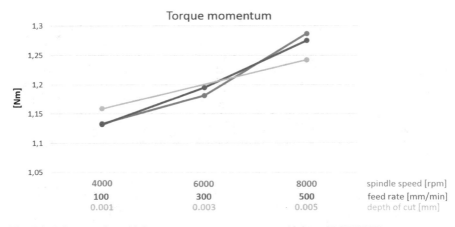

Fig. 7.4   Influence of machining parameters on torque at machining of PCBN [55]

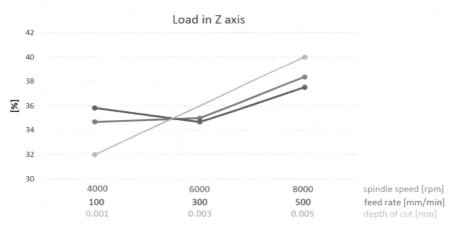

Fig. 7.5   Influence of machining parameters on load in Z-axis at machining of PCBN [55]

increasing of cutting speed caused rapid increasing of tool wear [55]. Surface roughness parameter Ra was in range from 0.15 to 0.30 μm depending on machining
parameters [61].

According to performed experiment, there is recommended following machining
parameters during RUM of PCBN to use: high depth of cut (5 μm) in terms of higher
tool life; other parameters depend on the size of the cutting tool. When rigid enough
cutting tool (e.g. Ø 24 mm) is used: medium cutting speed (450 m/min) in terms
of lower machine load; medium feed rate (300 mm/min) in terms of lower machine
load as well (machine load in Z-axis is the most restrictive parameter). However,
when small tool (e.g. Ø 2 mm) is used: low cutting speed (300 m/min) in terms of
lower torque; low feed rate (100 mm/min) in terms of lower torque as well (when

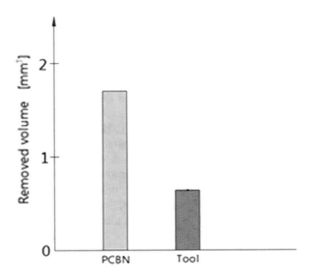

Fig. 7.6 Volume removed from PCBN and volume removed from the tool [55]

machining with smaller tools, torque lower than 1 Nm is recommended). The process was not as stable as the process during machining of MCD.

7.3 Aluminium Oxide

Corundum (alumina—Al_2O_3) is the second hardest mineral. It has value 9 on the Mohs scale of hardness [corresponding with the hardness up to 25 GPa (2500 HV)] [37]. It is considered as the hardest oxide [7]. The harder mineral is only diamond (Mohs hardness 10). Corundum has a hexagonal structure, and it is naturally transparent; however, impurities can change its colour [26]. For example, when chromium is present in corundum, it becomes red and the resultant mineral is called ruby. When titanium with iron is present, the resultant colour is blue and the mineral is called sapphire [21]. Synthetic corundum is usually white, and then, it is considered as a ceramic. It is manufactured from bauxite by Bayer process and calcining. Nevertheless, corundum consists of light elements (aluminium—2700 kg m^{-3} and oxygen—1.43 kg m^{-3}), its resultant density is relatively high—4020 kg m^{-3} and the melting temperature is 2072 °C [34, 71, 81].

Thanks to its properties (e.g. elastic modulus 375 GPa, coefficient of thermal expansion 8.4×10^{-6} K^{-1}), alumina offers a wide field of applications [70]. Alumina is usually used as abrasive (often on grinding wheels or sandpapers) [5, 36]. It finds application in lasers, satellites and spacecraft, as a scratch-resist optics, bearings, etc. It is an electrical insulator. Its thermal conductivity is about 30 W/(m K), which is relatively high for ceramic material (ceramics are often utilized as a thermal insulator) [2]. It found usage also in cosmetics, electronics, chemistry, dentistry, etc. [19, 66, 68].

There were performed much more experiments with rotary ultrasonic machining of alumina in comparison with the machining of MCD and PCBN. In comparison with the machining of MCD and PCBN (according to experiment condition used during machining of those materials), no significant loads were obtained during machining of aluminium oxide [55]. With increasing of cutting speed, the torque was increasing; however, other parameters were not affected, not even for changing of feed rate and depth of cut [59]. No significant tool wear was obtained. Altogether, it makes the ratio of material removed from a workpiece to the material removed from a tool (i.e. grinding ratio) approximately 70:1, as shown in Fig. 7.7.

With increasing spindle speed, the tool wear is slightly increasing. Other parameters had no significant effect on tool wear in the condition of experiments (cutting speed 300–600 m/min; feed rate 100–500 mm/min; DOC 1–5 μm). However, according to other experiments, there can be recommended machining parameters for the cutting tool Ø 24 mm: cutting speed 400 m/min; feed rate 1000 mm/min; and DOC 20 μm. Those parameters leave the machined surface, which does not need semi-finishing or finishing. However, for roughing is possible to use parameters: feed rate 50 mm/min with DOC 2500 μm (over 6 times higher MRR, but steps remain on the machined surface) [55].

The manufacturer of machine tools for rotary ultrasonic machining (DMG Mori, division Sauer) performed experiments with aluminium oxide as well. Their selected achievements during machining alumina ceramic are summarized in Tables 7.2 and 7.3. In these tables are recorded results of machining porous alumina for medical

Fig. 7.7 Volume removed from alumina and volume removed from the tool [55]

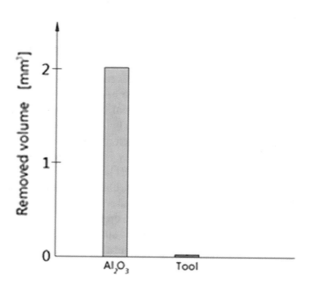

Table 7.2 Drilling of alumina components [92, 93, 122, 127]

Tool	Diameter (mm)	Cutting speed (m/min)	Feed rate (mm/min)	Infeed (mm)	Result
Drill	0.3	10	1.5	0.1	
Drill	1.2	30	10	0.1	
Drill	4	50	200	0.01	
Drill	8	100	5	0.2	

application [92], also contain the results of machining of part for X-ray machine [109] and other components made of alumina [93, 94, 122], 127.

According to experiments provided by Sauer division, which are summarized in Tables 7.2 and 7.3, there can be seen that cutting speed is varying in a large range. Feed rate is generally higher for the milling tools than the hollow drills. The bigger the tool is, the higher the infeed can be used (due to stiffness).

7.4   Zirconium Oxide

Zirconium dioxide (zirconia—ZrO_2) is a white ceramic material with very low thermal conductivity. It is chemically unreactive, and it has superior thermal, mechanical and electrical properties [143]. Zirconia is produced by calcining. By Kroll process (heating with carbon and chlorine), it is possible to obtain pure zirconium (Zr) [74]. Pure zirconia has a monoclinic crystal structure at room temperature; however,

Table 7.3 Milling of alumina components [92, 94, 109, 127]

Tool	Diameter (mm)	Cutting speed (m/min)	Feed rate (mm/min)	Infeed (mm)	Result
Mill	4	65	400	0.03	
Mill	5	95	500	0.05	
Mill	6	95	500	0.1	
Mill	20	350	1000	0.1	
Wheel	125	470	30	1.25	

when it is stabilized by yttrium it has a metastable tetragonal structure [27]. Presence of metastable phase enhances fracture toughness—when a higher amount of stress is present, the phase will change to the stable monoclinic phase, which absorbs part of the stress energy. This ability is called the transformation toughness [82]. Zirconium dioxide stabilized by yttrium is used especially in dental and medicine applications [69]. In dentistry, it is used to manufacture dental crowns and bridges, or full ceramic dental prosthesis [80, 140]. In aviation, it is used for thermally loaded components of diesel and jet engines [14].

Zirconia has thermal expansion similar to the steel (11×10^{-6} K^{-1}), low thermal conductivity (2.5 W/m K) and high fracture toughness (8 MPa m$^{1/2}$) [12]. The density of zirconia is 5680 kg/m^3 and its melting point is 2715 °C. It has high hardness

[12 GPa (1200 HV)], high wear resistance and good frictional behaviour [62]. Therefore, it can be used for valves, impellers, pump seals, oxygen sensors, grinding media, fuel cell membranes, medical prostheses, cutting blades, gears, bearings, etc. [83].

Wider application of the zirconia ceramic relates to more research and development of its production and processing. Sauer division of DMG Mori made several application experiments. Some of them, like manufacturing dental prosthesis [128], tooth crown [129], dental implant [95], dental bridge [123], dental bolt [124], zirconia part [125], zirconia disc [126], zirconia component [96], nozzle [110], etc., are summarized in Tables 7.4 and 7.5.

According to experiments provided by Sauer division, which are summarized in Tables 7.4 and 7.5, there can be seen that cutting speed is varying in a large range. Usually, there was used a lower cutting speed for smaller tools. It is due to the limitation of maximum spindle speed when ultrasound is active. Feed rate is generally higher for milling tools than hollow drills.

Table 7.4 Drilling of zirconia components [96, 110, 125, 128]

Tool	Diameter (mm)	Cutting speed (m/min)	Feed rate (mm/min)	Infeed (mm)	Result
Drill	0.42	13	0.9	0.01	
Drill	1.7	65	300	–	
Drill	6.25	60	4	0.1	
Drill	20	220	2000	0.01	

Table 7.5 Milling of zirconia components [95, 123, 124, 128, 129]

Tool	Diameter (mm)	Cutting speed (m/min)	Feed rate (mm/min)	Infeed (mm)	Result
Mill	1.7	45	200	0.01	
Mill	3	110	300	0.02	
Mill	5.2	160	1000	0.02	
Mill	8	200	1000	0.01	
Mill	24	600	3000	0.02	
Mill	33	410	800	0.09	

7.5   Silicon Carbide

Silicon carbide (carborundum—SiC) is considered as the hardest ceramics due to its high hardness 28 GPa (2800 HV). It has a black colour, high thermal and chemical stability, elastic modulus 410 GPa and the fracture toughness 4.6 MPa m$^{1/2}$. It has also low density ($3100\,kg/m^3$), high thermal conductivity ($120\,W/m\,K$) and low coefficient of thermal expansion (4×10^{-6} K^{-1}); therefore, it has great thermal shock resistance. SiC can be used at temperature up to 1650 °C [3]. It is used mostly as an abrasive [86]; however, due to its properties, it is also used in high-performance applications, such as turbine component, suction box covers, heat exchangers, valves, nozzles, seals, bearing, electronic component, furnaces, automotive components, bulletproof vests, etc. [16, 22]. SiC is produced by the electrochemical reaction of carbon (C) and sand (SiO_2) at high temperature (1600–2500 °C). It can be used as coating via chemical vapour deposition (CVD) [67]. Products can be made by sintering of SiC powder [31].

Sauer division made several application experiments with SiC ceramic. Some of them are recorded in Tables 7.6 and 7.7, like drilling experiments [130], sealing rings [131], grooving [133], lightweight core for mirror [98], milling experiments [111], SiC rings [112], hip joints [118], etc.

According to experiments provided by Sauer division, which are summarized in Tables 7.6 and 7.7, there can be seen that cutting speed is varying in a large range. Usually, there was used a lower cutting speed for smaller tools. It is due to the limitation of maximum spindle speed when ultrasound is active. Feed rate is generally higher for milling tools than hollow drills. The bigger the tool is, the higher the infeed can be used (due to stiffness).

Table 7.6 Drilling of carborundum components [130, 132, 134]

Tool	Diameter (mm)	Cutting speed (m/min)	Feed rate (mm/min)	Infeed (mm)	Result
Drill	1	8	1	0.05	
Drill	2.8	80	1.2	0.01	
Drill	5	65	5	0.05	

Table 7.7 Milling of carborundum components [97, 98, 111, 112, 131–134]

Tool	Diameter (mm)	Cutting speed (m/min)	Feed rate (mm/min)	Infeed (mm)	Result
Mill	0.4	8	30	0.001	
Mill	1.7	32	800	0.005	
Mill	5	90	500	0.005	
Mill	9	240	200	0.01	
Mill	16	270	1000	0.01	
Mill	30	360	1500	0.02	
Mill	50	70	500	0.025	

(continued)

Table 7.7 (continued)

Tool	Diameter (mm)	Cutting speed (m/min)	Feed rate (mm/min)	Infeed (mm)	Result
Mill	90	710	1	10	
Mill	150	565	10	10.5	

7.6   Other Materials

Other materials suitable for rotary ultrasonic machining include materials, such as aluminium nitride (AlN), silicon nitride (Si_3N_4), borosilicate glass [$SiO_2 + B_2O_3$ (in Table 7.8 labelled as SiB)], silicon oxide (SiO_2), silicon (Si), boron carbide (B_4C), ytterbium oxide (Yb_2O_3), sapphire, ruby, onyx, glass (like BK7), etc. Those materials are typical by their high hardness, which causes problems during machining

Table 7.8 Drilling of other materials

Material	Tool	Diameter (mm)	Cutting speed (m/min)	Feed rate (mm/min)	Infeed (mm)	References
AlN	Drill	6	90	5	0.1	[119]
Si_3N_4	Drill	2.6	65	1.2	0.01	[135]
Si_3N_4	Drill	6.8	85	0.5	0.01	[136]
Si_3N_4	Drill	9	57	3	0.5	[138]
Si_3N_4	Drill	0.6	11	50	0.005	[99]
SiB	Drill	4.6	22	20	1	[102]
SiB	Drill	0.4	10	2.5	0.2	[103]
SiO_2	Drill	33	156	5	0.5	[139]
Si	Drill	3	38	1000	0.02	[88]
Si	Drill	8	100	5	0.1	[89]
Si	Drill	0.5	13	25	0.25	[90]
B_4C	Drill	1.9	24	1	0.01	[106]
BK7	Drill	7	66	30	2	[115]
BK7	Drill	2	38	20	0.01	[117]

by conventional technologies. Generally, they are kinds of ceramics, minerals, gems, glasses. Some examples of machining parameters provided by Sauer division are recorded in Tables 7.8 and 7.9.

According to experiments provided by Sauer division, which are summarized in Tables 7.8 and 7.9, there can be seen that cutting speed is varying in a large range. It is caused by different mechanical properties of machined material. However, machining parameters are varying even for the same material type—it is caused by the application. Different machining parameters were used for roughing and for finishing. Size and shape of the workpiece and dimensions and shape of the machined feature also affect machining parameters. Usually, there was used a lower cutting speed for smaller tools. Feed rate is generally higher for milling tools than hollow drills. The bigger the tool is, the higher the infeed can be used.

Beside above-mentioned hard and brittle materials, rotary ultrasonic machining can be used for machining of materials which can be hardened by cold forming.

Table 7.9 Milling of other materials

Material	Tool	Diameter (mm)	Cutting speed (m/min)	Feed rate (mm/min)	Infeed (mm)	References
Si_3N_4	Mill	33	360	2000	0.05	[137]
Si_3N_4	Mill	80	754	100	0.02	[100]
Si_3N_4	Mill	24	400	200	0.005	[101]
Si_3N_4	Mill	1.3	24	400	0.005	[120]
Si_3N_4	Mill	8	133	1500	0.03	[113]
Si_3N_4	Mill	3	50	1200	0.06	[121]
SiO_2	Mill	90	565	0.1	3.8	[139]
Si	Mill	90	707	500	0.15	[87]
Si	Mill	33	332	1000	0.25	[104]
B_4C	Mill	90	850	300	0.005	[105]
B_4C	Mill	2	25	250	0.005	[106]
Yb_2O_3	Mill	30	283	400	0.01	[114]
Yb_2O_3	Mill	5	50	600	0.02	[114]
BK7	Mill	40	628	800	0.1	[107]
BK7	Mill	50	545	1000	0.25	[91]
BK7	Mill	100	1095	500	0.005	[91]
BK7	Mill	80	628	500	0.12	[108]
BK7	Mill	4	57	960	0.063	[115]
BK7	Mill	1.3	20	1000	0.01	[116]
BK7	Mill	10.6	100	24	1	[116]
BK7	Mill	6	106	600	0.1	[117]

And with combination with ultrasonic-assisted machining, there can be machined also materials like chromium-cobalt (Co–Cr), titanium (Ti), wolfram carbide (WC), graphite (C), martensite (Fe_3C), ferrite, vanadium (V), stainless steel (SS), composites reinforced by aramid/carbon/glass fibres (AFRP, CFRP, GFRP) and many more. Such a range of machinable materials makes rotary ultrasonic machining attractive in sectors, where machining of different kinds of unique materials is needed.

References

1. 3B_diamond (2019) Changsha 3 better ultra-hard materials, Co. Ltd. http://www.3bdiamond. com/userlist/xiongxl/text-710.html. Accessed 21 July 2019
2. Accuratus (2014) Aluminum oxide, Al_2O_3 ceramic properties. http://accuratus.com/alumox. html. Accessed 25 July 2014
3. Accuratus (2018) Silicon carbide, SiC ceramic properties. https://accuratus.com/silicar.html. Accessed 12 Dec 2018
4. Ahmed W, Sein H, Ali N, Gracio J, Woodwards R (2003) Diamond films grown on cemented WC-Co dental burs using an improved CVD method. Diam Relat Mater 12(8):1300
5. Alumina (2014) Alumina (aluminium oxide)—the different types of commercially available grades. The A to Z of Materials
6. Anthony JW, Bideaux RA, Bladh KW, Nichols MC (1990) Graphite. Handbook of mineralogy. I (elements, sulfides, sulfosalts). Mineralogical Society of America, Chantilly, VA, USA. ISBN 0962209708
7. Anthony JW, Bideaux RA, Bladh KW, Nichols MC (1997) Corundum. Handbook of mineralogy III (halides, hydroxides, oxides). Mineralogical Society of America, Chantilly, VA, USA. ISBN 0962209724
8. Békés J, Hrubec J, Kicko J, Lipa Z (1999) Teória obrábania. Publisher STU, Bratislava, 157 p. ISBN 80-227-1261-2
9. Blank V, Popov M, Pivovarov G, Lvova N, Gogolinsky K, Reshetov V (1998) Ultrahard and superhard phases of fullerite C60: comparison with diamond on hardness and wear. Diam Relat Mater 7(2–5):427
10. Calculator (2014) Calculator for conversion between vickers hardness number and SI units MPa and GPa. http://www.gordonengland.co.uk/hardness/hvconv.htm. Accessed 18 July 2014
11. Catledge SA, Vohra YK (1999) Effect of nitrogen addition on the microstructure and mechanical properties of diamond films grown using high-methane concentrations. J Appl Phys 86:698
12. Ceramic industry (2018) Oxide ceramics—zirconium oxide. The all-purpose construction material. https://www.ceramtec.com/ceramic-materials/zirconium-oxide/. Accessed 2 Oct 2012
13. Chih-Shiue Y, Ho-Kwang M, Wei L, Jiang Q, Yusheng Z, Russell JH (2005) Ultrahard diamond single crystals from chemical vapor deposition. Physica Status Solidi (a) 201(4):R25. https:// doi.org/10.1002/pssa.200409033
14. Clarke DR, Oechsner M, Padture NP (2012) Thermal-barrier coatings for more efficient gas-turbine engines. Cambridge University Press 37(10):890–941
15. Coelho RT, Yamada S, Aspinwall DK, Wise MLH (1995) The application of polycrystalline diamond (PCD) tool materials when drilling and reaming aluminum-based alloys including MMC. Int J Mach Tools Manuf 35(5):761
16. Davis RF (2017) Silicon carbide. Reference module in materials science and materials engineering
17. Delhaes P (2001) Graphite and precursors. CRC Press. ISBN 90-5699-228-7

18. Denisenko A, Kohn E (2005) Diamond power devices. Concepts and limits. Diam Relat Mater 14(3–7):491
19. Dentallab (2014) Modern dental laboratory USA—Procera alumina. http://moderndentalusa.com/products/all-ceramic/procera-alumina/. Accessed 26 July 2014
20. Dey SR, Meshram MP, Kodli BK (2014) Friction stir welding of austenitic stainless steel by PCBN tool and its joint analyses. In: Procedia materials science, vol 6. 3rd International conference on materials processing and characterization (ICMPC), pp 135–139
21. Duroc-Danner JM (2011) Untreated yellowish orange sapphire exhibiting its natural color. J Gemmol 32:175–178
22. Ebnesajjad S (2014) Surface treatment and bonding of ceramics. Surface treatment of materials for adhesive bonding, pp 283–299
23. Ekimov EA, Sidorov VA, Bauer ED, Mel'nik NN, Curro NJ, Thompson JD, Stishov SM (2004) Superconductivity in diamond. Nature 428(6982):542–545. arXiv:cond-mat/0404156
24. Fischer-Cripps AC (2004) Nanoindentation. Springer, New York, p 198. ISBN 0-387-22045-3
25. Fukunaga O (2002) Science and technology in the recent development of boron nitride materials. J Phys Condens Matter 14(44):10979
26. Gray T (2009) The elements. Illinois, USA, 240 pp. ISBN 978-80-7391-544-5
27. Greenwood NN, Earnshaw A (1997) Chemistry of the elements. Butterworth-Heinemann, Oxford. ISBN 0-7506-3365-4
28. Greim J, Schwetz KA (2005) Boron carbide, boron nitride, and metal borides. Ullmann's encyclopedia of industrial chemistry. Wiley-VCH, Weinheim
29. Guides (2008) Guides for the jewelry, precious metals, and pewter industries. 2008. Federal trade commission letter declining to amend the guides with respect to use of the term "cultured", U.S. Federal Trade Commission, July 21
30. HPHT (2014) High pressure, high temperature. International diamond laboratories. http://www.diamondlab.org/80-hpht_synthesis.htm. Accessed 20 July 2014
31. Hansson T, Warren R (2000) Particle and whisker reinforced brittle matrix composites. Compr Compos Mater 4:579–609
32. Harris DC (1999) Materials for infrared windows and domes: properties and performance. SPIE Press, pp 303–334. ISBN 0-8194-3482-5
33. Heartwig J et al (2006) Diamonds for modern synchrotron radiation sources. European Synchrotron Radiation Facility
34. Higashi GS, Fleming CG (1989) Sequential surface chemical reaction limited growth of high quality Al_2O_3 dielectrics. Appl Phys Lett 55(19):1963–1965
35. Holtzapffel C (1856) Turning and mechanical manipulation. Holtzapffel, pp. 176–178. ISBN 1-879335-39-5
36. Hudson LK, Misra Ch, Perrotta AJ, Wefers K, Williams FS (2002) Aluminum oxide. In Ullmann's encyclopedia of industrial chemistry. Wiley-VCH, Weinheim
37. Hurlbut CS, Klein C (1985) Manual of mineralogy, 20th edn. Wiley, pp 300–302. ISBN 0-471-80580-7
38. Ito E, Schubert G (2007) Multianvil cells and high-pressure experimental methods. In: Treatise of geophysics 2. Elsevier, Amsterdam, pp. 197–230. ISBN 0-8129-2275-1
39. Jackson DD, Aracne-Ruddle C, Malba V, Weir ST, Catledge SA, Vohra YK (2003) Magnetic susceptibility measurements at high pressure using designer diamond anvils. Rev Sci Instrum 74(4):2467
40. Janáč A, Kicko J, Lipa Z, Charbula J, Peterka J (1994) Technológia obrábania, montáže a základy strojárskej metrológie. Publisher STU, Bratislava, 317 p. ISBN 80-227-0698-1
41. Janáč A, Lipa Z, Charbula J, Peterka J, Görög A (2002) Technológia obrábania a metrológia. Publisher STU, Bratislava, 194 p. ISBN 80-227-1711-8
42. Janáč A, Bátora B, Baránek I, Lipa Z (2004) Technológia obrábania. Publisher STU, Bratislava, 289 p. ISBN 80-277-2031-3
43. Janáč A, Lipa Z, Peterka J (2006) Teória obrábania. Publisher STU, Bratislava, 199 p. ISBN 80-227-2347-9

44. John P, Polwart N, Troupe CE, Wilson JIB (2002) The oxidation of (100) textured diamond. Diam Relat Mater 11(3–6):861
45. Kawaguchi M et al (2008) Electronic Structure and Intercalation chemistry of graphite-like layered material with a composition of BC_6N. J Phys Chem Solids 69(5–6):1171
46. Khakani MA, Chaker M (1993) Physical properties of the X-ray membrane materials. J Vac Sci Technol B 11(6):2930–2937
47. Kim (2014) Do Kyung Kim. Nanoindentation lecture 1 basic principle. Department of Material Science and Engineering, KAIST, Korea
48. Komatsu T et al (1999) Creation of superhard B-C–N heterodiamond using an advanced shock wave compression technology. J Mater Process Technol 85:69
49. Kupec T, Hlaváčová I, Turňa M (2012) Friction stir welding of Mg and Al alloys. In: Advanced materials research. International conference on advanced material and manufacturing science (ICAMMS), vol 875–877. Trans Tech Publications, China, Beijing, pp 1477–1482, 6 p
50. Kupec T, Turňa M, Zifčák P (2013) Friction Stir welding of magnesium alloy type AZ 61. In: IN-TECH 2013: Proceedings of international conference on innovative technologies, Budapest, Hungary—Rijeka: Faculty of Engineering University of Rijeka, pp 153–156. ISBN 978-953-6326-88-4
51. Kupec T, Turňa M, Kuruc M, Jáňa M, Behúlová M (2014) Influence of tool geometry on the quality of aluminum alloy weld joints produced by FSW. In: Proceedings of the 10th international symposium on friction stir welding. TWl Ltd., Beijing, China—Cambridge, 10 p. ISBN 978-1-903761-10-6
52. Kuruc M, Peterka J (2014a) Rotary ultrasonic machining of poly-crystalline cubic boron nitride. In: IDS 2014. International doctoral seminar: proceedings of the 9th international doctoral seminar, Zielona Góra, Poland. – Zielona Góra: University of Zielona Góra, pp 110–116. ISBN 978-80-8096-195-4
53. Kuruc M, Necpal M, Peterka J (2014b) Machining of poly-crystalline cubic boron nitride by laser beam machining in terms of surface roughness. J Prod Eng 17(1):101–104. ISSN 1821-4932
54. Kuruc M, Vopát T, Peterka J (2014c) Surface roughness of poly-crystalline cubic boron nitride after rotary ultrasonic machining. In: Annals of DAAAM and Proceedings of DAAAM Symposium: Collection of Working Papers for 25th DAAAM international symposium 25(1). ISSN 2304–1382
55. Kuruc M (2015a) Ultrasonic machining. Dissertation thesis, STU MTF, 172 pp
56. Kuruc M, Peterka J (2015b) Influence of machining parameters on machine tool loads at rotary ultrasonic machining of synthetic diamond. In Proceedings of TEAM 2015: 7th international scientific and expert conference of the international TEAM society, Belgrade, Serbia. Belgrade: Faculty of Mechanical Engineering, pp 598–601. ISBN 978-86-7083-877-2
57. Kuruc M, Vagovský J, Peterka J (2015c) Influence of machining parameters on machine tool loads at rotary ultrasonic machining of cubic boron nitride. In ICPM 2015: Proceedings of the 8th international congress on precision machining. Novi Sad, Serbia. Novi Sad: Faculty of Technical Sciences, pp 195–200. ISBN 978-86-7892-742-3
58. Kuruc M, Kusý M, Šimna V, Peterka J (2015d) Influence of machining parameters on surface topography of cubic boron nitride at rotary ultrasonic machining. In ICPM 2015: Proceedings of the 8th international congress on precision machining. Novi Sad, Serbia. Novi Sad: Faculty of Technical Sciences, pp 157–162. ISBN 978-86-7892-742-3
59. Kuruc M (2015e) Machine tool loads in rotary ultrasonic machining of alumina, CBN and synthetic diamond. In: Proceedings of the 26th DAAAM international symposium. DAAAM International, Viedeň, pp 519–523. ISSN 1726-9679. ISBN 978-3-902734-07-5
60. Kuruc M, Necpal M, Jáňa M, Peterka J (2015f) Comparison of machining of poly-crystalline cubic boron nitride by rotary ultrasonic machining and laser beam machining in terms of chemical affection. In: Automation in production planning and manufacturing: 16th international scientific conference for Ph.D. students. Žilina, pp 69–74. ISBN 978-80-89276-47-9
61. Kuruc M, Urminský J, Necpal M, Morovič L, Peterka J (2015g) Comparison of machining of poly-crystalline cubic boron nitride by rotary ultrasonic machining and laser beam machining

in terms of shape geometry. In: Strojírenská technologie – Plzeň: 6th International Conference, Plzeň, ČR. Plzeň: Západočeská univerzita v Plzni, pp 122–128. ISBN 978-80-261-0304-2

62. Kyocera (2018). Kyocera global. Fine ceramics (advanced ceramics). Zirconia – technical data. https://global.kyocera.com/prdct/fc/list/material/zirconia/zirconia.html. Accessed 15 Dec 2018

63. Lanhua W, Kuo P, Thomas R, Anthony T, Banholzer W (1993) Thermal conductivity of isotopically modified single crystal diamond. Phys Rev Lett 70(24):3764–3767

64. Lavrakas V (1957) The lubricating properties of graphite. J Chem Educ 34(5):240

65. Leichtfried G et al (2002) Properties of diamond and cubic boron nitride. In: Landolt-Börnstein—group VIII, advanced materials and technologies: powder metallurgy data. Refractory, Hard and Intermetallic Materials 2A2. Springer, Berlin, pp 118–139

66. Levin I, Brandon D (1998) Metastable alumina polymorphs: crystal structures and transition sequnces. J Am Ceram Soc 81(8):1995–2012

67. Locke CW, Severino A, Via FV, Reyes M, Register J, Saddow SE (2012) SiC films and coatings. Silicon Carbide Biotechnology, pp 17–61

68. Mallick PK (2008) Fiber-reinforced composites materials, manufacturing, and design, 3rd ed (Ch. 2.1.7). CRC Press, Boca Raton, FL. ISBN 0-8493-4205-8

69. Manicone PF, Iommetti PR, Raffaelli L (2007) An overview of zirconia ceramics: basic properties and clinical applications. J Dent 35(11):819–826

70. Material (2014) Material properties data: alumina (aluminum oxide). http://www.makeitfrom.com. Accessed 24 July 2014

71. Nanodiamond (2014) Nanodiamond and superhard thin-films. http://www.cityu.edu.hk/cos daf/cbn_property.htm. Accessed 17 July 2014

72. Nassau K, Nassau J (1979) The history and present status of synthetic diamond. J Cryst Growth 46(2):157

73. Neves AJ, Nazaré MH (2001) Properties, growth and applications of diamond. IET, pp 142–147. ISBN 0-85296-785-3

74. Nielsen R (2005) Zirconium and zirconium compounds. In: Ullmann's encyclopedia of industrial chemistry. Wiley-VCH, Weinheim. https://doi.org/10.1002/14356007.a28_543

75. Nusinovich GS (2004) Introduction to the physics of gyrotrons. JHU Press, p 229. ISBN 0-8018-7921-3

76. Oliver WC, Pharr GM (1992) An improved technique for determining hardness and elastic modulus using load and displacement sensing indentation experiments. J Mater Res 7(6)

77. Osawa E (2007) Recent progress and perspectives in single-digit nanodiamond. Diam Relat Mater 16(12):2018–2022

78. Pan Z et al (2009) Harder than diamond: superior indentation strength of Wurtzite BN and lonsdaleite. Phys Rev Lett 102(5):055503

79. Panizza M, Cerisola G (2005) Application of diamond electrodes to electrochemical processes. Electrochim Acta 51(2):191

80. Papaspyridakos P, Kunal L (2008) Complete arch implant rehabilitation using subtractive rapid prototyping and porcelain fused to zirconia prosthesis: a clinical report. J Prosthet Dent 100(3):165–172. https://doi.org/10.1016/S0022-3913(08)00110-8

81. Patnaik P (2002) Handbook of inorganic chemicals. McGraw-Hill. ISBN 0-07-049439-8

82. Porter DL, Evans AG, Heuer AH (1979) Transformation toughening in PSZ. Acta Metall 27:1649. https://doi.org/10.1016/0001-6160(79)90046-4

83. Precision Ceramics (2019) Advanced materials solution. https://www.precision-ceramics.co.uk/. Accessed 10 Dec 2019

84. Railkar TA, Kang WP, Windischmann H, Malshe AP, Naseem HA, Davidson JL, Brown WD (2000) A critical review of chemical vapor-deposited (CVD) diamond for electronic applications. Crit Rev Solid State Mater Sci 25(3):163

85. Read PG (2005) Gemmology. Butterworth-Heinemann, pp 49–50. ISBN 0-7506-6449-5

86. Rowe WB (2014) Grinding wheel developments. Principles of Modern Grinding Technology, pp 35–62 (2014)

87. Sauer report (2004n) Report #20040103. Monokristallines Silizium. Ergebnisse der Muster-bearbeitung
88. Sauer report (2004o) Report #20040170. Lightweight Silicon Core for Mirror. Results of test trial
89. Sauer report (2004p) Report #20040185. Si circular blank. Results of test trial
90. Sauer report (2004r) Report #20040213. Silicium Rechteckplatte. Ergebnisse der Musterbear-beitung
91. Sauer report (2004s) Report #20040073. BK7 lens. Results of test trial
92. Sauer report (2005a) Report #20050183. Medical implant. Results of test trial
93. Sauer report (2005b) Report #20050095. Alumina ceramic. Results of test trial
94. Sauer report (2005c) Report #20050133. Alumina component. Results of test trial
95. Sauer report (2005d) Report #20050112. Dental implant. Results of test trial
96. Sauer report (2005e) Report #20050049. Zirconia component. Results of test trial
97. Sauer report (2005f) Report #20050118. SiC component. Results of test trial
98. Sauer report (2005g) Report #20050158. Lightweight SiC core for mirror. Results of test trial
99. Sauer report (2005h) Report #20050075. CMM Probes. Results of test trial
100. Sauer report (2005i) Report #20050080. Prägeform. Ergebnisse der Musterbearbeitung
101. Sauer report (2005j) Report #20050170. Umformwerkzeug. Ergebnisse der Musterbear-beitung
102. Sauer report (2005k) Report #20050137. Borosilikat Glasspritzen zur Dosierung kleinster Flüssigkeitsmengen. Ergebnisse der Musterbearbeitung
103. Sauer report (2005l) Report #20050147. Glasspritzen zur Dosierung kleinster Flüssigkeits-mengen. Ergebnisse der Musterbearbeitung
104. Sauer report (2005m) Report #20050016. Silicium Rechteckblock. Ergebnisse der Muster-bearbeitung
105. Sauer report (2005n) Report #20050070. Bohrkarbid Rohteil. Ergebnisse der Musterbear-beitung
106. Sauer report (2005o) Report #20050071. B4C Rohteil. Ergebnisse der Musterbearbeitung
107. Sauer report (2005p) Report #20050011. Leiste „Halter unten". Ergebnisse der Musterbear-beitung
108. Sauer report (2005r) Report #20050029. Optical and lasers prism. Ergebnisse der Muster-bearbeitung
109. Sauer report (2006a) Report #20060106. X-ray machine part. Results of test trial
110. Sauer report (2006b) Report #20060102. Nozzle. Technology report
111. Sauer report (2006c) Report #20060095. SiSiC block. Technology report
112. Sauer report (2006d) Report #20060097. SiC ring. Technology report
113. Sauer report (2006e) Report #20060078. Verkleidung für Schmelzöfen. Technology report
114. Sauer report (2006f) Report #20060031. Ytterbium Oxide. Technology report
115. Sauer report (2006g) Report #20060028. BK7 glass block. Technology report
116. Sauer report (2006h) Report #20060079. BK7 tube. Technology report
117. Sauer report (2006i) Report #20060082. BK7 block. Technology report
118. Sauer report (2007a) Report #20070046. Hip Joint. Technology report
119. Sauer report (2007b) Report #20071120. AlN Testteil
120. Sauer report (2007c) Report #20070005. Si_3N_4 test pieces. Technology report
121. Sauer report (2007d) Report #20070036. Gehäuse. Technology report
122. Sauer report (2008a) Report #20080001. Al^2O^3 test part
123. Sauer report (2008b) Report #20080002. Zirconium Tooth Bridge
124. Sauer report (2008c) Report #20080019. Zirconium Oxide Bolt
125. Sauer report (2008d) Report #20080020. Zirconium implant
126. Sauer report (2003) Report #20030039. Zirconia disc. Results of sample processing
127. Sauer report (2004a) Report #20040179. Alumina part. Results of test trial
128. Sauer report (2004b) Report #20040160. Zirconia prosthesis. Results of sample processing
129. Sauer report (2004c) Report #20040162. Tooth crown. Results of test trial
130. Sauer report (2004d) Report #20040001. Drilling of SiC. Results of test trial

131. Sauer report (2004e) Report #20040058. Sealing rings. Results of test trial
132. Sauer report (2004f) Report #20040064. Probeplatte. Results of test trial
133. Sauer report (2004g) Report #20040207. Test material. Results of test trial
134. Sauer report (2004h) Report #20040096. Ring. Results of test trial
135. Sauer report (2004i) Report #20040043. Bohren von Kanälen. Ergebnisse der Musterbearbeitung
136. Sauer report (2004j) Report #20040068. Bohrung. Ergebnisse der Musterbearbeitung
137. Sauer report (2004k) Report #20040134. Heißpresswerkzeug. Ergebnisse der Musterbearbeitung
138. Sauer report (2004l) Report #20040199. Drilling hole to SiN. Results of test trial
139. Sauer report (2004m) Report #20040002. Silikat Glaslinse. Ergebnisse der Musterbearbeitung
140. SheN J (2013) Advanced ceramics for dentistry. Elsevier/BH, Amsterdam, 271 pp. ISBN 978-0123946195
141. Silberberg MS (2009) The molecular nature of matter and change. In: Chemistry, 5th edn. McGraw-Hill, New York, p 483. ISBN 0073048593
142. Smither RK, Davey S, Purohit A (1992) Diamond monochromator for high heat flux synchrotron X-ray beams. In: Khounsary AM (ed) Proceedings of the SPIE, vol 1739. High Heat Flux Engineering, pp 628–642
143. Stevens R (1986) Introduction to Zirconia. In: Magnesium Elektron Publication. No 113
144. Sumiya H (2005) Super-hard diamond indenter prepared from high-purity synthetic diamond crystal. Rev Sci Instrum 76(2):026112
145. Supply (2013) The state of 2013 global rough diamond supply. Resource Investor. 22 Jan 2013
146. Taniguchi T et al (2002) Ultraviolet light emission from self-organized p–n domains in cubic boron nitride bulk single crystals grown under high pressure. Appl Phys Lett 81(22):4145
147. Todd RH, Allen DK, Alting L (1994) Manufacturing processes reference guide. Industrial Press Inc., pp 43–48. ISBN 0-8311-3049-0
148. UHD (2018) Ultra-hard China. https://www.ultrahard-china.com/large-single-crystal-diamond/large-single-crystal-diamond.html. Accessed 1 Dec 2018
149. Unidens (2014) zubná technika. http://unidens.com/web/website/mainmenu/mainpage/. Accessed 19 July 2014
150. Vel L et al (1991) Cubic boron nitride: synthesis, physicochemical properties and applications. Mater Sci Eng, B 10(2):149
151. Veprek S, Zeer A, Riedel R (2000) In: Riedel R (ed) Handbook of ceramic hard materials. Wiley, Weinheim 2000. ISBN 3-527-29972-6
152. Wentorf RH (1961) Synthesis of the cubic form of boron nitride. J Chem Phys 34(3):809–812
153. Wentorf RH, Devries RC, Bundy FP (1980) Sintered superhard materials. Science 208(4446):873–880. https://doi.org/10.1126/science.208.4446.873. PMID 17772811
154. Werner M, Locher R (1998) Growth and application of undoped and doped diamond films. Rep Prog Phys 61(12):1665
155. Yongjun T et al (2013) Ultrahard nanotwinned cubic boron nitride. Nature 493(7432):385–388
156. Zoski CG (2007) Handbook of electrochemistry. Elsevier, p 136. ISBN 0-444-51958-0

Chapter 8
Application of Rotary Ultrasonic Machining for Cutting-Edge Preparation

Abstract Cutting-edge preparation brings a lot of benefits in the machining process, including increased tool life, process stability and surface quality. Therefore, there were developed several methods to fulfil this edge modification. Those methods differ by principle, application, effectivity and costs. However, create cutting-edge preparation into CBN inserts remains challenging. Those inserts are used for heavy-duty applications, such as machining of hardened steels and cast irons. Their main characteristic is very high hardness, which makes them abrasive resistant. But there is one machining method created to processing hard materials—rotary ultrasonic machining (RUM). This chapter describes and analyses the experiment of application of RUM to produce cutting-edge preparation on CBN inserts. Experiments confirmed the suitability of this process—it successfully creates a chamfer. However, the accuracy was disputable. There are determined possible reasons for inaccuracy and suggested options of solution for increase the precision of achieved cutting-edge preparation.

Cutting-edge microgeometry affects many aspects of a machining operation, such as cutting force [7], quality of the machined surface, mechanical stress [2] and temperature distribution on the cutting tool, as well as tool wear and tool life [16]. Sharp cutting edge contains defects and tool tracks from the previous processing. Moreover, sharp edge has low strength and bend resistance. Such low mass can be rapidly overheated in cutting conditions. Therefore, sharp cutting tool is relatively unstable with short tool life. Moreover, relatively low quality of sharp edge surface decreases adhesiveness of the coatings, which decrease tool life even more [17]. Those issues can be solved by controllable blunting—cutting-edge preparation, as shows Fig. 8.1.

Cutting-edge preparation is one of the methods to increase the durability of the cutting tool [20]. It also increases adhesiveness of the coatings. The principle of this process is to create small fillet or chamfer on the cutting edge. Examples of the shapes of prepared edges are shown in Fig. 8.2.

For this purpose, the following methods are mostly used: brushing (using nylon brushes impregnated with abrasive grits) [18]; microblasting (using abrasive jet—a high-pressured air stream to propel small abrasive particles) [12]; drag finishing (the tools rotate in liquid polishing process media) [6]; abrasive flow machining (extruding

© The Author(s), under exclusive license to Springer Nature Switzerland AG 2021
M. Kuruc, *Rotary Ultrasonic Machining*, Manufacturing and Surface Engineering,
https://doi.org/10.1007/978-3-030-67944-6_8

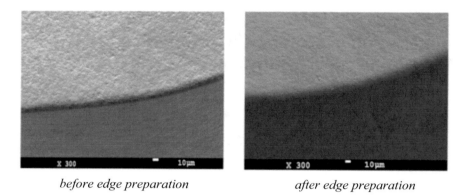

before edge preparation   *after edge preparation*

Fig. 8.1 Cutting-edge microgeometry observed by electron microscope [21]

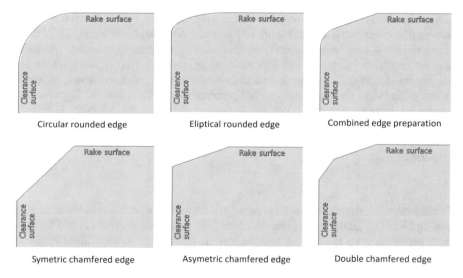

Fig. 8.2 Typical geometries of the prepared cutting edges

an abrasive paste around the cutting edge) [2]; magnetic abrasive machining (using a magnetic field to generate polishing forces for abrasive ferromagnetic grains) [13]; laser marking (using ablation to removing layers of material) [1]; electro-erosion honing (using electrical discharge machining via sinking of counterface electrode) [22].

Most of the above-mentioned methods (brushing, dry blasting, wet blasting, drag finishing, abrasive flow machining, magnetic abrasive machining) are utilizing an abrasive effect for cutting-edge preparation. It means, their application and productivity are affecter and limited by hardness of cutting edge. Considering CBN cutting inserts, those methods are not suitable for their processing. The laser can effectively machine CBN. However, the setting of proper machining parameters is very difficult. Moreover, early melting of some phases with lower melting temperature (e.g. binder) can occur [11]. A necessary condition for electro-erosion is the electrical conductivity of the substrate, and CBN has very low electrical conductivity [5]. According to results of rotary ultrasonic machining of cubic boron nitride, this method seems to be suitable for processing CBN cutting insets. Considering the RUM process characteristics, chamfered edge is easier to create. Rounded edge is also possible (due to five axes kinematic); however, the machining time would be much longer (and possibly lesser accurate).

8.1 Experimental Set-Up

The aim of the experiment is to determine if rotary ultrasonic machining (RUM) is possible to achieve accuracy enough for the cutting-edge preparation—due to very low dimensions of this microgeometry, it is challenging to reach specific values of rounding or chamfering. Even small difference in its value can cause different behaviours, as shown Fig. 8.3. There were compared different fillets of the cutting-edge during machining of steel X12CrMoVNbN9-1 at cutting speed 200 m min^{-1}, feed per tooth 4.5 mm and depth of cut 0.02 mm. Cutting-edge radius is labelled as r_β and B means volume of removed material [11].

From Fig. 8.3, it can be seen that the higher radius of the cutting edge is, the longer tool life, lower surface roughness and lower cutting force are achieved. The recommended value of the rounding is 20–25 μm. The above-mentioned experiment would not exceed depth of cut (20 μm); therefore, the highest value of radius was 15 μm [11]. Most of the cutting-edge preparation processes are using the energy source (mechanical, thermal, …) for the creation of cutting-edge preparation. When the substrate is exposed for a longer time to this energy, the bigger cutting-edge preparation is created. With increasing the demanded size of cutting-edge preparation, the processing time is drastically increased (due to the bigger amount of removed mass). Therefore, there has to be found a compromise between size of the radius and economy of the process (due to energy and time consumption). However, the RUM process works differently. The whole cutting-edge preparation can be created by a

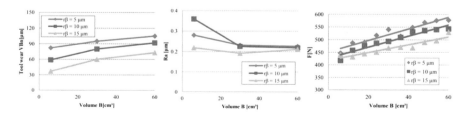

Fig. 8.3 Influence of the cutting-edge radius on tool wear, surface roughness and cutting force [11]

single tool path. The size of the cutting edge preparation affects processing time only a little.

To determine RUM suitability to the process of the cutting-edge preparation, there were created simple chamfering on the CBN samples. The chamfering differs between samples by size and angle. Initial cutting insert has dimensions $12.7 \times 12.7 \times 4.8$ mm and already contains chamfer 0.185 mm $\times 21°$ with rounding 0.03 mm at clearance surface. This initial cutting-edge preparation with its microstructures can be shown in Fig. 8.4. The rake face is presented as nearly horizontal surface, and clearance face is presented as nearly vertical surface. Between them can be seen the area of the cutting-edge preparation. The magnification of the microstructures is approximately $5000\times$.

Due to initial cutting-edge preparation, there were selected for experiments sizes of chamfering: 0.2 mm, 0.3 mm, 0.4 mm and 0.5 mm and angles of chamfering: 15°, 20°, 25°, 30° and 35°. Scheme of this edge preparation is shown in Fig. 8.5.

There were made eight experiments. Required values are recorded in Table 8.1. Labelling of parameters is based on Fig. 8.5. Important values are a (angle) and b (length). There were calculated parameters x (distance in X-axis) and y (distance in Y-axis) due to easier evaluations. And for programming of CNC machine tool,

Fig. 8.4 Magnified initial cutting-edge preparation and its microstructures

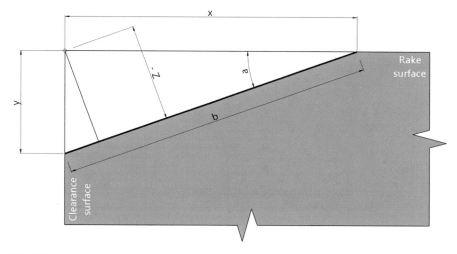

Fig. 8.5 Labelling of the cutting-edge parameters

Table 8.1 Required values of cutting-edge parameters

Experiment	a (°)	b (mm)	x (mm)	y (mm)	$-Z$ (mm)
#1	20	0.212836	0.2	0.072794	0.068404
#2	20	0.319253	0.3	0.109191	0.102606
#3	20	0.425671	0.4	0.145588	0.136808
#4	20	0.532089	0.5	0.181985	0.171010
#5	15	0.517638	0.5	0.133975	0.129410
#6	25	0.551689	0.5	0.233154	0.211309
#7	30	0.577350	0.5	0.288675	0.250000
#8	35	0.610387	0.5	0.350104	0.286788

there were computed also value $-Z$, which determine depth, which is needed to reach required values (3 + 2 machining were used).

DMG Ultrasonic 20 linear was used as machine tool. It is machine tool designed for rotary ultrasonic machining (RUM) with five controllable axes. Accurate positioning ($\pm2.5\,\mu$m) is provided by linear motors, which allows using feed rate 40 m min^{-1} with acceleration 2 g [8]. Mounted spindle can reach spindle speed up to 42,000 rpm. Ultrasonic energy (with frequency 20–30 kHz) can be transferred via this spindle. Tool magazine has 24 slots, including touch probe and calibre. In front of tool magazine is laser probe for precise measuring of the cutting tools [15].

Diamond ultrasonic milling cutter was used as cutting tool. There were chosen the cup-wheel cutter (6A9) with an outer diameter of 24 mm [10]. This ultrasonic mill looks like grinding wheel, but it works at much lower cutting speeds in comparison with grinding process. The tool is hollow due to internal cooling [19]. Used machining

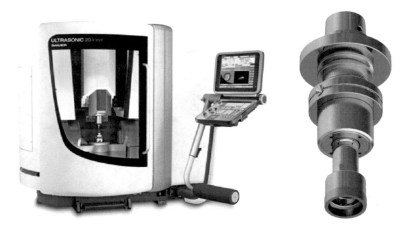

Fig. 8.6 Machine tool and cutting tool used in the experiments [14]

parameters were: cutting speed 450 m min^{-1} (spindle speed was 5968 rpm), feed rate 300 mm min^{-1} (feed per revolution was 0.05 mm) and depth of cut 0.005 mm. The used machine tool and milling cutter are shown in Fig. 8.6.

Before every machining, there was measured the position of the sample, including its inclination. For every edge was remeasured the coordinate system (zero point). Even tool length was remeasured. This procedure allows the best possible accuracy. After machining of all eight edges on the CBN cutting insert, the actual dimensions were measured by Zeiss Surfcom 5000.

8.2 Achieved Results

Zeiss Surfcom 5000 is a measuring machine for roughness and contour measurements [3]. The precision of the measurement is ±0.2 μm. The device is flexible and easy to automatize. It finds application in automotive, mechanical and medical industries. It is suitable for measuring lenses, bearings, spindles, injections and many other mostly grinded or lapped parts [9]. The measured sample and measuring device are shown in Fig. 8.7.

Results of the measurement are recorded in the following figures. Every prepared edge was measured three times—near both corners and in the middle. Idea was to determine the inclination if it is present there. Figure 8.8 shows the result of the first edge at 500× magnification. Due to space saving and readability, only one image was selected. Obtained values are summarized in Table 8.2. The labelling of the parameters is based on Fig. 8.5. There are recorded values from all three positions and their average value was calculated. Then there was calculated the deviation as the difference between required value and average value from measurements. There was calculated a percentage deviation considering different sizes of the measured parameters. To determine the wideness of the interval, there was calculated the variance as well.

Fig. 8.7   Prepared CBN insert and measuring machine [4]

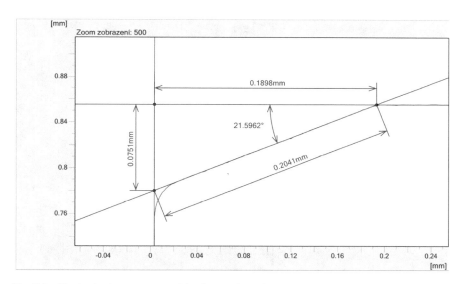

Fig. 8.8   Obtained microgeometry of the first cutting edge

Figure 8.8 shows a radius at clearance surface. This radius was not created during the experiments, which means that there was not reached enough depth. It means the machining was not present here. According to Table 8.2, it can be concluded that angle deviation is the highest—it is caused by the positioning precision of the cradle. Cradle and rotating table motors do not reach accuracy of the linear motors for translation movements. However, there is also achieved the lowest value of the

Table 8.2 Required, measured and calculated values for the first prepared cutting edge

Edge A/parameter	Required value	Position 1	Position 2	Position 3	Average value	Deviation	Percentage deviation	Variance
a (°)	20	21.4525	21.5962	21.5781	21.5423	1.5423	7.7113	0.004083629
b (μm)	212.8	202.8	204.1	203.3	203.4	−9.4	−4.4333	0.286666667
x (μm)	200.0	188.7	189.8	189.1	189.2	−10.8	−5.4000	0.206666667
y (μm)	72.8	74.2	75.1	74.8	74.7	1.9	2.6183	0.140000000

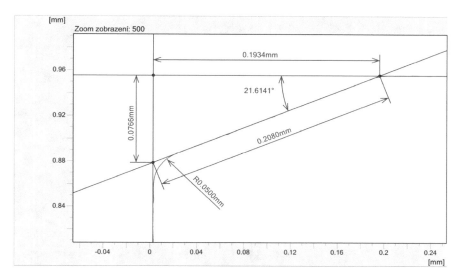

Fig. 8.9 Obtained microgeometry of the second cutting edge

variance—it is caused by strong brake for the cradle and low cutting force in the process (caused by low material removal rate).

Figure 8.9 shows the result of the second edge at 500× magnification. The parameters were measured at three positions on the cutting edge, but due to space saving and readability, also, only one image was selected. Obtained values are summarized in Table 8.3. The labelling of the parameters is based on Fig. 8.5.

The radius at clearance surface can be obtained even in this case, as shown Fig. 8.9. Probably, no machining was present here as well. According to Table 8.3, it can be concluded that the deviations of linear characteristics are much higher in comparison with the previous experiment. It is caused by lack of tool contact—required values were increased, but achieved ones were almost the same.

Figure 8.10 shows the result of the third edge at the same magnification. The parameters were measured at three positions on the cutting edge, but due to space saving and readability, also, only one image was selected. Obtained values are summarized in Table 8.4. The labelling of the parameters is based on Fig. 8.5.

In Fig. 8.10, the radius at clearance surface cannot be obtained. It means that machining was present here this time. According to Table 8.4, it can be also concluded that the deviations of linear characteristics are much higher in comparison with the first experiment. However, they are very similar with deviations of the second experiment. The samples were measured after machining, which means, if there were some error of tool length measuring in the machine tool, this error will be repeated for all edges.

Figure 8.11 shows the result of the fourth edge. This edge has bigger size than all previous; therefore, magnification 200× was used. The parameters were measured at three positions on the cutting edge, but due to space saving and readability, also, only

Table 8.3 Required, measured and calculated values for the second prepared cutting edge

Edge B/parameter	Required value	Position 1	Position 2	Position 3	Average value	Deviation	Percentage deviation	Variance
a (°)	20	21.6141	21.6339	21.5819	21.6010	1.6100	8.050	0.000459209
b (μm)	319.3	208.0	208.1	208.5	208.2	−111.1	−34.7853	0.046666667
x (μm)	300.0	193.4	193.4	193.9	193.6	−106.4	−35.4778	0.055555556
y (μm)	109.2	76.6	76.7	76.7	76.7	−32.5	−29.7867	0.002222222

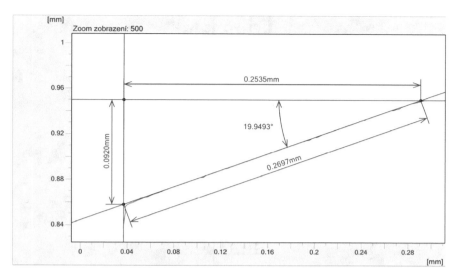

Fig. 8.10 Obtained microgeometry of the third cutting edge

one image was selected. Obtained values are summarized in Table 8.5. The labelling of the parameters is based on Fig. 8.5.

In Fig. 8.11, also, the radius at clearance surface cannot be obtained. Machining was present here as well. According to Table 8.5, it can be also concluded that the deviations of linear characteristics are lower in comparison with the second and third experiments. The tool length error in measurement is lowering with increasing of machined depth—therefore, percentage deviation has approximately a half value in comparison with the third experiment.

Those four experiments (the first group of experiments) were made on the top side of the inserts. The next four experiments (the second group of experiments) were made on the bottom side. Using the same specimen should prevent against affection of minor difference in chemical and mechanical properties between other inserts. For this second group of experiments, an angle was changing, and linear parameter x was constant—its value was 0.5 mm to ensure contact of the cutting tool with the workpiece. This specimen and its labelling of the edges are shown in Fig. 8.12. There can be seen that edge E is under edge A, edge F under edge D, edge G under edge C and edge H under edge B.

Figure 8.13 shows the result of the fifth edge. This edge has similar size as the previous one (fourth); therefore, the same magnification of $200\times$ was used. The parameters were measured at three positions on the cutting edge, but due to space saving and readability, also, only one image was selected. Obtained values are summarized in Table 8.6. The smallest angle was used—it required the biggest inclination of the cradle. Using of even smaller angle could cause a collision of the tool-holder with the vice. The labelling of the parameters in Table 8.6 is based on Fig. 8.5.

Table 8.4 Required, measured and calculated values for the third prepared cutting edge

Edge C/parameter	Required value	Position 1	Position 2	Position 3	Average value	Deviation	Percentage deviation	Variance
a (°)	20	19.9493	20.0448	19.9564	19.9835	−0.0165	−0.0825	0.001887
b (μm)	425.7	269.7	268.4	268.3	268.8	−156.9	−36.8527	0.406667
x (μm)	400.0	253.5	252.2	252.2	252.6	−147.4	−36.8417	0.375556
y (μm)	145.6	92.0	92.0	91.6	91.9	−53.7	−36.8996	0.035556

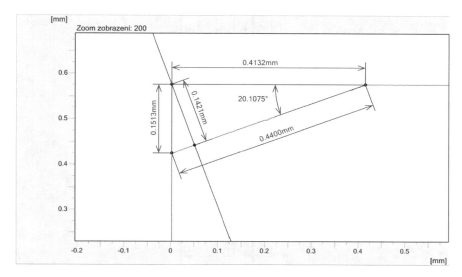

Fig. 8.11 Obtained microgeometry of the fourth cutting edge

In Fig. 8.13 the radius at clearance surface is not present. Machining was present here as well. According to Table 8.6, it can be also concluded that the deviations of linear characteristics are similar to the ones of the fourth experiment. Variance is significantly lower in comparison with the previous (fourth) experiment, which could be caused by lower achieved value of the observed parameters.

Figure 8.14 shows the result of the sixth edge. This edge has also similar size as the previous one; therefore, the same magnification was used. The parameters were measured at three positions on the cutting edge, but due to space saving and readability, there were selected also only one image. Obtained values are summarized in Table 8.7. The labelling of the parameters is based on Fig. 8.5.

In Fig. 8.14 the radius at clearance surface is not present. Machining was present here as well. According to Table 8.7, it can be concluded that the deviations of linear characteristics are almost double in comparison with the fifth experiment, almost the same to the third experiment. Variance is similar to the previous (fifth) experiment. Achieved angular and linear characteristics are smaller than required values. It could be caused by positioning for rotary axis (for angle) and tool wear (for distance).

Figure 8.15 shows the result of the seventh edge. This edge has also similar size as the previous one (only angle was changing); therefore, the same magnification was used. The parameters were measured at three positions on the cutting edge, but due to space saving and readability, also only one image was selected. Obtained values are summarized in Table 8.8. The labelling of the parameters is based on Fig. 8.5.

In Fig. 8.15 the radius at clearance surface is not present as well (machining was present here). According to Table 8.8, it can be concluded that the deviations of linear characteristics are lower in comparison with the sixth experiment. Variance is significantly lower than in the previous (sixth) experiment.

Table 8.5 Required, measured and calculated values for the fourth prepared cutting edge

Edge D/parameter	Required value	Position 1	Position 2	Position 3	Average value	Deviation	Percentage deviation	Variance
a (°)	20	20.1075	19.9400	19.9941	20.01387	0.01387	0.0693	0.004871
b (μm)	532.1	440.0	434.4	428.4	434.3	−97.8	−18.3846	22.43556
x (μm)	500.0	413.2	408.4	402.6	408.1	−91.9	−18.3867	18.78222
y (μm)	181.0	151.3	148.2	146.5	148.7	−33.3	−18.3083	3.948889

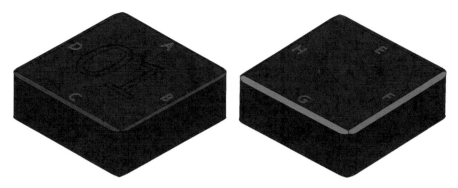

Fig. 8.12 Illustration of the top and bottom side of the specimen with labelled edges

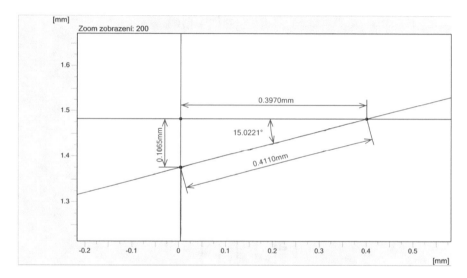

Fig. 8.13 Obtained microgeometry of the fifth cutting edge

Figure 8.16 shows the result of the last (eighth) edge. The same magnification was used as in previous causes—200×. The parameters were measured at three positions on the cutting edge, but due to space saving and readability, also, only one image was selected. Obtained values are summarized in Table 8.9. The labelling of the parameters is based on Fig. 8.5.

In Fig. 8.16 the radius at clearance surface is not present as well (machining was present here). According to Table 8.9, it can be concluded that the deviations of linear characteristics are almost the same in comparison with the seventh experiment. Variance is significantly higher than in the previous (seventh) experiment.

Table 8.6 Required, measured and calculated values for the fifth prepared cutting edge

Edge E/parameter	Required value	Position 1	Position 2	Position 3	Average value	Deviation	Percentage deviation	Variance
a (°)	15	15.0835	14.9718	15.0221	15.0258	0.0258	0.1720	0.002086
b (μm)	517.6	408.3	409.3	411.0	409.5	−108.1	−20.8842	1.242222
x (μm)	500.0	394.2	395.4	397.0	395.5	−104.5	−20.8933	1.315556
y (μm)	134.0	106.2	105.7	106.5	106.1	−27.8	−20.7810	0.108889

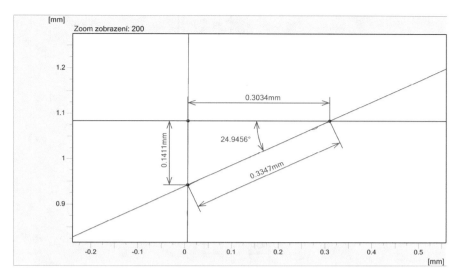

Fig. 8.14 Obtained microgeometry of the sixth cutting edge

In all cases, the variance of the angle was always the lowest. It is caused by strong brake on the cradle and low cutting forces. It means, even if there was some deviation caused by accuracy of the positioning, this position was stable for the whole machining. Deviation of the linear characteristics was significantly higher. It could be caused by tool wear—CBN is very abrasive material, and even there was not machined big amount of the material, considering small dimensions of required geometry even such small tool wear can affect the results. Another reason for such deviation could be an error in the laser probe calibration. This probe was measuring the cutting tool before and after machining of each edge; however, difference in tool length does not respond to deviation.

The deviations of angle are shown in Fig. 8.17. Source data comes from tables 8.2 to 8.9. Angle 20° was used four times; however, only the last one was taken into consideration, because only there was confirmed machining of whole surface. Every value is a result of arithmetical value of three measurements. There can be seen that there is no direct connection between deviation and angle.

The deviations of distance x are shown in Fig. 8.18. Source data comes from tables 8.2 to 8.9. There was used distance 0.5 mm for five times, and therefore, its value was calculated as arithmetical value of all those distances (arithmetical value of 15 values). However, it is important to mention that at distance 0.2 and 0.3 mm, machining of the sample was not observed; therefore, their informative value is low.

The experiment confirmed that rotary ultrasonic machining (RUM) can be used for machining of CBN cutting inserts. The cutting-edge preparation was achieved; however, due to its small size, it could be challenging to reach proper accuracy. The precision of the angle was in range −0.09° to +0.04°, which can be considered as accurate enough. However, the average deviation of the distance was −0.134 mm for

Table 8.7 Required, measured and calculated values for the sixth prepared cutting edge

Edge F/parameter	Required value	Position 1	Position 2	Position 3	Average value	Deviation	Percentage deviation	Variance
a (°)	25	24.9316	24.8514	24.9456	24.90953	−0.0905	−0.3619	0.001722
b (μm)	551.7	337.8	336.6	334.7	336.4	−215.3	−39.0297	1.628889
x (μm)	500.0	306.3	305.4	303.4	305.0	−195.0	−38.9933	1.468889
y (μm)	233.2	142.4	141.4	141.1	141.6	−91.5	−39.2533	0.308889

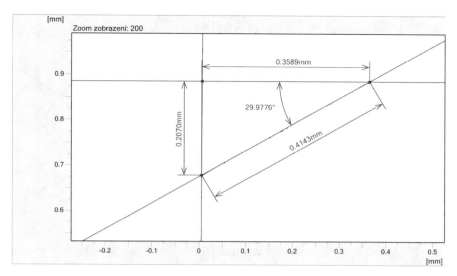

Fig. 8.15 Obtained microgeometry of the seventh cutting edge

parameter x 0.5 mm. The worst deviation at this parameter was at angle 25° (its value was −0.195 mm). At this same angle, also, the worst angular deviation was achieved. If there was considered only the angle 20° (as at 0.2 mm; 0.3 mm; and 0.4 mm), the worst deviation for 0.5 mm would be only −0.092 mm. This value corresponds with tool length difference 0.031 mm. If this value would be manually used to shorten the cutting tool in the tool register of the machine tool control system, the cutting tool will be sent to higher depth—deep enough to rapidly reduce the deviation. By then, achievable precision of the dimension would be considered as accurate enough. This value of the tool length correction can be calculated via the following formulas. To determine correction length, experimental machining is needed, which provides values for comparison. The labelling of the parameters is corresponding with Fig. 8.5.

$$\text{If } a \text{ and } b \text{ are known:} \quad x = \cos a \cdot b \tag{8.1}$$

$$\text{Then:} \quad -Z = \sin a \cdot x \tag{8.2}$$

$$\text{Combining the previous formulas:} \quad -Z = \sin a \cdot \cos a \cdot b \tag{8.3}$$

$$\text{After simplification:} \quad -Z = \frac{\sin 2 \cdot a}{2} \cdot b \tag{8.4}$$

Table 8.8 Required, measured and calculated values for the seventh prepared cutting edge

Edge G/parameter	Required value	Position 1	Position 2	Position 3	Average value	Deviation	Percentage deviation	Variance
a (°)	30	29.9778	29.9050	30.0186	29.9671	−0.03287	−0.1096	0.002208
b (μm)	577.4	414.3	413.7	413.2	413.7	−163.6	−28.3393	0.202222
x (μm)	500.0	358.9	358.7	357.8	358.5	−141.5	−28.3067	0.228889
y (μm)	288.7	207.0	206.3	206.7	206.7	−82.0	−28.4086	0.082222

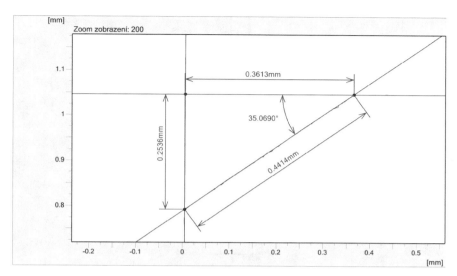

Fig. 8.16 Obtained microgeometry of the eighth cutting edge

If a and x are known, only Eq. (8.2) can be used. If $-Z$ is known, no formula is needed. If any different combinations of parameters are known, basic goniometry is enough to calculate required parameters. Parameter $-Z$ determines depth of the cutting tool. If achieved chamfer is smaller than required one, tool did not reach demanded depth. Then, $-Z$ for required chamfer and $-Z$ for achieved chamfer are needed to calculate. Difference between those two values is value, which is needed to deduct from the tool length in machine tool control system. Then, machine tool will send cutting tool deeper—deep enough to reach requested depth.

When applying Eq. (8.4), the difference in depth for used experiments can be calculated. The results are summarized in Table 8.10.

Table 8.10 shows relatively high variance of values. Results of the first two edges (A and B) are irrelevant, because they were not machined. All the time, the achieved depth was lower than required. Average value of the difference is 57 μm. Higher difference is present at higher angles of chamfer.

Cutting tool length was measured before every machining. During measuring process, the tool is measured several times, and the result value is a mean value of all measurements. There is just one condition—the variance of values has to be smaller than 0.01 mm. Otherwise, the measurement is considered as unsuccessful. However, the tool length can be slightly different during machining process—when the ultrasound is active, the tool vibrates, which causes periodical length changes with size of ultrasonic amplitude, which has in this case value about 10 μm. Spindle temperature can also affect the length (due to thermal expansion). Moreover, tool wear is present. There were subtracted only small amount of material; however, CBN is very abrasive material, and even such small amount of material can wear the tool. However, the tool should not lose more than few micrometres. The tool depth is calculated form coordinate system. This system has to be measured by touch probe first. The precision of the probe is affected by calibration, but its value should be

Table 8.9 Required, measured and calculated values for the eight prepared cutting edge

Edge H/parameter	Required value	Position 1	Position 2	Position 3	Average value	Deviation	Percentage deviation	Variance
a (°)	35	35.0690	35.0560	34.9969	35.04063	0.0406	0.1161	0.000984
b (μm)	610.4	441.4	438.3	442.4	440.7	−169.7	−27.7999	3.046667
x (μm)	500.0	361.3	358.8	362.4	360.8	−139.2	−27.8333	2.268889
y (μm)	350.1	253.6	251.7	253.7	253.0	−97.1	−27.7357	0.846667

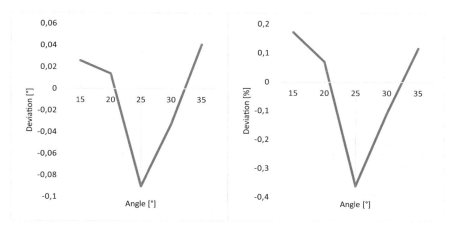

Fig. 8.17 Influence of angle deviation on angle of cradle

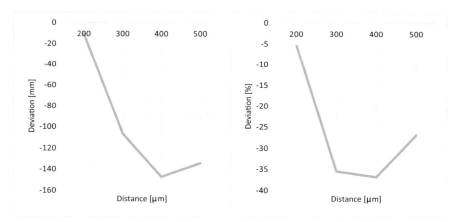

Fig. 8.18 Influence of selected distance deviation on size of chamfer

	Edge	$-Z_{required}$ (μm)	$-Z_{achieved}$ (μm)	Depth difference (μm)
Table 8.10 Required, measured and calculated values for the seventh prepared cutting edge	A	68.404	69.472	-1.070
	B	102.606	71.288	31.318
	C	136.808	86.337	50.471
	D	171.010	139.660	31.350
	E	129.410	102.544	26.866
	F	211.309	128.476	82.833
	G	250.000	179.055	70.945
	H	286.788	207.175	79.613

lesser than 5 μm. Toughness of the system also should not make more than a few micrometres. Combination of those deviation origins together can cause issues with precision. Anyway, there can be concluded the possible sources of inaccuracy to:

- precision of cutting tool measurement,
- size of ultrasonic amplitude,
- warming up the spindle before measurement,
- tool wear,
- precision of coordinate system determination,
- toughness of the system,
- condition of machine tool.

In terms of increasing the accuracy of the process, it is recommended to remeasure cutting tool before every machining; remeasure coordinate system after every reclamping of the cutting insert (even when clamping jig is used); remeasure inclination of the clamped cutting insert. And after obtaining of the inaccuracy, calculation of the error and its consecutive compensation in NC programme should increase the precision even more.

The microstructure of created cutting-edge preparations in terms of possible damaging of the surface of the CBN cutting inserts was also analysed. The results are summarized in Fig. 8.19. Magnification of approximately 5000× was used.

According to observed microstructures, there can be seen the cracks in the CBN grains. Only the edge A seems without cracks; however, according to measured dimensions of the chamfer, there were not expected machining at all.

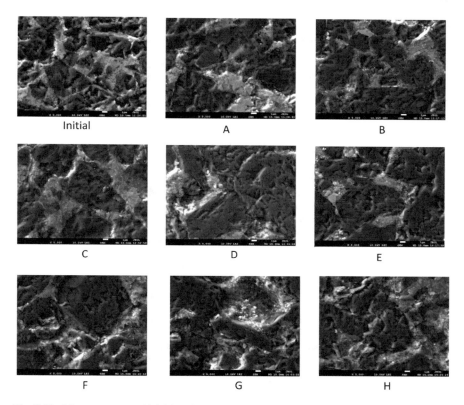

Fig. 8.19 Microstructures of initial and created chamfers

References

1. Aurich JC, Zimmermann M, Leitz L (2011) The preparation of cutting edges using a marking lasers. Prod Eng Res Devel 5(1):17–24
2. Bouzakis K-D, Michailidis N, Skordaris G, Kombogiannis S, Efstathiou K, Erkens G, Rambadt S, Wirth I (2002) Effect of the cutting edge radius and its manufacturing procedure on the milling performance of PVD coated cemented carbide inserts. CIRP Ann Manuf Technol 51:61–64
3. Carl Zeiss (2020a) Contour and surface measuring machines. [cit. 08. 08. 2020]. Available at: https://www.zeiss.com/metrology/products/systems/form-and-surface/contour-measurement.html. Accessed 8 Aug 2020
4. Carl Zeiss (2020b) Shape measuring machine Surfcom 5000. https://www.directindustry.com/prod/carl-zeiss-industrielle-messtechnik-gmbh/product-5693-1071113.html. Accessed 20 Mar 2020
5. Crookall JR, Fereday RJ (1973) An experimental determination of the degeneration of tool-electrode shape in electro-discharge machining. Microtecnic 17:197–200
6. Cselle T (2007) Influence of edge preparation on the performance of coated cutting tools. In: International conference on metallurgical coatings and thin films. 34 p
7. Denkena B, Kohler J, Mengasha MS (2012) Influence of the cutting edge rounding on the chip formation process: Part 1. Investigation of material flow, process forces, and cutting temperature. In: Production engineering, vol 6, Issue 4–5, pp 329–338

8. DirectIndustry (2020a) Ultrasonic machining center with linear motors. http://www.direct industry.com/prod/dmg-mori/ultrasonic-machining-centers-linear-motors-5973-555063.html. Accessed 22 Feb 2020

9. Engineering (2020) Surface measuring system—surfcom 5000. https://datasheets.globalspec. com/ds/4248/CarlZeissIMT/1ADDE0CC-4852-4723-B397-F763C823CDF7. Accessed 2 Feb 2020

10. FEPA (2019) Federation of European producers of abrasives. http://www.fepa-abrasives.org/. Accessed 1 Aug 2019

11. Fulemova J, Janda Z (2014) Influence of the cutting edge radius and the cutting edge preparation on tool life and cutting forces at inserts with wiper geometry. Proc Eng 69:565–573

12. Chastagner MW, Shih A (2007) Abrasive jet machining for edge generation. Trans NAMRI/SME 35:359–366

13. Cheung FY, Zhou ZF, Geddam A, Li KY (2008) Cutting edge preparation using magnetic polishing and its influence on the performance of high-speed steel drills. J Mater Process Technol 208(1–3):196–204

14. Kuruc M (2015) Ultrasonic machining. Dissertation thesis. STU MTF, 172 p

15. PD HSC (2009) Planning documentation HSC/ultrasonic 20 linear. Germany, 102 p

16. Rech J, Yen Y-C, Schaff MJ, Hamdi H, Altan T, Bouzakis KD (2005) Influence of cutting edge radius on the wear resistance of PM-HSS milling inserts. Wear 259(7–12):1168–1176

17. Rodriguez CJC (2009) Cutting edge preparation of cutting tools by applying micro-abrasive jet machining and brushing. Kassel University Press, Germany, 189 p. ISBN: 978-3-89958-712-8

18. Shaffer WR (1999) Cutting tool edge preparation. In: Technical paper—society of manufac turing engineers, pp 1–8

19. Schott (2013) Diamond tools: ultrasonic. http://www.schott-diamantwerkzeuge.com/ultras onic.html. Accessed 1 Sept 2013

20. Vopát T (2015) The influence of edge preparation on the tool life of coated cutting tools. Dissertation thesis, STU MTF, 134 p

21. Vopát T, Podhorský Š, Sahul M, Haršáni M (2019) Cutting edge preparation of cutting tools using plasma discharges in electrolyte. J Manuf Process 46:234–240

22. Yussefian NZ, Koshy P, Buchholz S, Klocke F (2010) Electro-erosion edge honing of cutting tools. In: CIRP annals—manufacturing technology, vol 59, Issue 1, pp 215–218

Chapter 9
Conclusion

Ultrasonic energy can positively affect many processes. One of them is machining. During rotary ultrasonic machining and ultrasonic-assisted milling, the ultrasound causes:

- a decrease in cutting force (~50%),
- a decrease in heat generation (~40%),
- decreasing of surface roughness parameters (Ra ~ 0.1 μm),
- allows the machining of hard and brittle materials (HV ~ 100 GPa).

The machining parameters (e.g. cutting speed, feed rate, depth of cut) have a great influence on the resultant quality of machined surfaces, wear of ultrasonic tool, loads of machine tool, grinding ratio, material removal rate, etc. The self-sharping effect is often typical for the process of rotary ultrasonic machining. This effect is based on replacing worn abrasive particles with the new ones, which are present in the volume of the active part of the ultrasonic tool. Due to the lack of optimal machining parameters for rotary ultrasonic machining of very hard materials, recommended proper machining parameters for selected hard materials in consideration of relevant outputs according to performed researches are suggested. For example, there can be recommended for rotary ultrasonic face milling by ultrasonic cutter with diameter 24 mm following the machining parameters:

- ZrO_2—$v_c = 600$ m min^{-1}; $v_f = 1000$ mm min^{-1}; $a_p = 0.020$ mm.
- Al_2O_3—$v_c = 400$ m min^{-1}; $v_f = 1000$ mm min^{-1}; $a_p = 0.020$ mm.
- SiC—$v_c = 350$ m min^{-1}; $v_f = 1000$ mm min^{-1}; $a_p = 0.010$ mm.
- CBN—$v_c = 400$ m min^{-1}; $v_f = 300$ mm min^{-1}; $a_p = 0.005$ mm.
- MCD—$v_c = 500$ m min^{-1}; $v_f = 100$ mm min^{-1}; $a_p = 0.005$ mm.

Rotary ultrasonic machining (RUM) can be applied: in sectors focused especially on manufacturing or modification of tools (especially cutting tools) made of very hard materials; in electrical engineering or jewellery (processing of semiconductors and gems); in stomatology and medicine (production of prosthesis); in mechanical engineering (production of fine mechanics, dies and moulds); as well as in aerospace

© The Author(s), under exclusive license to Springer Nature Switzerland AG 2021
M. Kuruc, *Rotary Ultrasonic Machining*, Manufacturing and Surface Engineering, https://doi.org/10.1007/978-3-030-67944-6_9

industry and in other sectors, where only advanced materials with superior properties are demanded. Experiments have proved that there is no limitation in the hardness of a workpiece material for rotary ultrasonic machining. Therefore, it seems suitable for manufacturing the tools made of very hard materials, or for their cutting-edge preparation.

Cutting-edge preparation is a process of controlled "blunting" of the cutting tool. As a result, this process is increasing the tool life and stability of the machining process. For the most difficult application, cubic boron nitride (CBN) inserts are used. However, due to its mechanical properties (very high hardness), it is a challenge to process such a material. Most of the methods for cutting-edge preparation based on mechanical subtraction are ineffective. However, RUM is able to machine CBN effectively. Therefore, it was used as processing method for creating required chamfer on the cutting edge on CBN insert. Experiments confirmed suitability of this process—it successfully creates a chamfer. However, the accuracy was disputable, because average deviation of the angle was 4% and average deviation of dimension was 27%. Possible reasons of inaccuracy and suggested options of solution for increasing the precision of achieved cutting-edge preparation were determined.

In further experiments, these options and different cutting strategies can be applied. Instead of creating every chamfer separately, cutting strategy which creates all four chamfers on the single side of the inserts can be used, using an ultrasonic ball mill and CAM software. This option allows to make easily different chamfers and fillets and their combinations. Or continuous five axes machining for increased quality of the chamfered surface can be used.

Printed in the United States
By Bookmasters